Lecture Notes
in Control and Information Sciences   261

Editors: M. Thoma · M. Morari

Springer
London
Berlin
Heidelberg
New York
Barcelona
Hong Kong
Milan
Paris
Singapore
Tokyo

H.A. Talebi, R.V. Patel and K. Khorasani

Control of Flexible-link Manipulators Using Neural Networks

With 55 Figures

Springer

Authors

H.A. Talebi, PhD
Amirkabir University, Tehran, Iran

R.V. Patel, PhD
K. Khorasani, PhD
University of Western Ontario, London Ontario, Canada N6A 5B9

ISBN 1-85233-409-6 Springer-Verlag London Berlin Heidelberg

British Library Cataloguing in Publication Data
Talebi, H. A.
    Control of flexible-link manipulators using neural
    networks. - (Lecture notes in control and information
    sciences ; 261)
    1.Manipulators (Mechanism) - Automatic control. 2.Neural
    networks (Computer science)
    I.Title II.Patel, Rajnikant V. III.Khorasani, K.
    629.8'92
    ISBN 1852334096

Library of Congress Cataloging-in-Publication Data
A catalog record for this book is available from the Library of Congress

Typesetting: Camera ready by authors
Printed and bound at the Athenæum Press Ltd., Gateshead, Tyne & Wear
69/3830-543210 Printed on acid-free paper SPIN 10776687

To Sarah and Sahar (HAT)
To Roshni (RVP)
To Kamyar, Arianne, and Anahita (KK)

Preface

The problem of modeling and control of flexible–link manipulators has received much attention since the 1980's. There are a number of potential advantages stemming from the use of light–weight flexible–link manipulators, such as faster operation, lower energy consumption, and higher load–carrying capacity for the amount of energy expended. However, structural flexibility causes many difficulties in modeling the manipulator dynamics and guaranteeing stable and efficient motion of the manipulator end–effector. Control difficulties are mainly due to the non-colocated nature of the sensor and actuator positions, which results in non–minimum phase behavior, i.e., unstable zero dynamics. Further complications arise because of the highly nonlinear nature of the system and the difficulty involved in accurately modeling various friction and backlash terms. Control strategies that ignore these problems generally fail to provide satisfactory closed–loop performance.

A number of conventional linear as well as nonlinear techniques have been developed in recent years to address the problem of controlling flexible-link manipulators. However, the highly nonlinear nature of the manipulator dynamics and the inherent uncertainty associated with obtaining accurate dynamic models of flexible-link manipulators make reliable practical control of such manipulators difficult. In recent years, a number of intelligent control techniques, especially those based on artificial neural networks have been proposed for problems where the system dynamics are difficult to model or prone to uncertainties. A flexible-link manipulator is an example of such a system, and its control involving various neural network–based control strategies is the subject of this monograph. More specifically, the monograph presents experimental evaluation of the performance of neural network–based controllers for tip position tracking of flexible-link manipulators. The controllers are designed by utilizing the output redefinition approach to overcome the problem caused by the non–minimum phase characteristic of the flexible–link system. Four different neural network schemes are proposed. The first two schemes are developed by using a modified version of the "feedback–error–learning" approach to learn the inverse dynamics of the flexible-link manipulator. The neural networks are trained and employed as online controllers. Both schemes require *only* a linear model of the system for defining the new outputs and for designing conventional PD–type controllers. This assumption is relaxed in the third and fourth schemes. In the third scheme, the controller is designed based on tracking the hub position while controlling the elastic deflection at the tip. In the fourth scheme which employs two neural networks, the first network (referred to as the output neural network) is responsible for specifying an appropriate output for ensuring minimum phase behavior of the system. The second neural network is responsible for implementing an inverse dynamics controller. Both networks are trained online. Finally, the four proposed neural network controllers are implemented on a single flexible–link experimental test–bed. To improve the transient as well as steady-state response of the system, the two schemes are modified by adding a joint PD controller to the neural network controller. Experimental results show that this modification results in significant improvement in the system response and increases the dynamic range of the neural network controller and the robustness of the system to changes in the desired trajectory. Experimental and simulation results are presented to illustrate the advantages and improved performance of the proposed tip position tracking controllers over the conventional PD–type controllers in the presence of unmodeled dynamics such as hub friction and stiction and payload variations.

The funding for much of the research described in this monograph was provided by the Natural Sciences and Engineering Research Council (NSERC) of Canada and by Fonds pour la Formation de Chercheurs et l'Aide à la Recherche (FCAR) of the Province of Quebec. This support is gratefully acknowledged.

H.A. Talebi
R.V. Patel
K. Khorasani

Contents

List of Figures

List of Tables

List of Symbols

A	Link cross–sectional area
E	Young's modulus
e	Position error
$\dot{e}$	Velocity error
$\ddot{e}$	Acceleration error
F_1	Viscous damping at the hub
F_2	Damping matrix due to the internal viscous friction
f_c	Coulomb friction
H	Inverse of the inertia matrix
I	Area moment of inertia of the link
I_h	Hub inertia
J	Objective function of the neural network
J_0	Link inertia relative to the hub
J_l	Payload inertia
K	Stiffness matrix
K_p	Proportional gain of the PD controller
K_v	Velocity gain of the PD controller
l_i	Length of the ith link
M	Bending moment
$M(\theta, \delta)$	Inertia matrix
M_l	Payload mass
S	Shear force
u	Input torque
u_c	Output of the conventional controller
u_n	Output of the neural network controller
$W_i(x,t)$	Deflection of a point x along the ith link
$W_i(l_i,t)$	Tip deflection of the ith link

$\mathbf{w}$	Matrix of the weights of the neural network
y_{ai}	Redefined output for the ith link
y_r	Reference position trajectory
$\dot{y}_r$	Reference velocity trajectory
$\ddot{y}_r$	Reference acceleration trajectory
$y_r^{(r)}$	rth derivative of y_r
y_{ri}	Reflected tip position of the ith link
y_{ti}	Net tip position of the ith link
α_i	A scale factor corresponding to a specific point on the ith link
α_i^*	Critical value of α_i
α_{i0}	The value of α_i^* for zero payload mass
δ_{ij}	jth flexible mode of the ith link
η	Learning rate
θ_i	Joint position of the ith link
$\dot{\theta}_i$	Joint velocity of the ith link
$\ddot{\theta}_i$	Joint acceleration of the ith link
ρ	Uniform density of the link
τ_i	Input torque for the ith link
ϕ_{ij}	jth eigenfunction of the ith link
ω_{ij}	jth natural frequency of the ith link

Chapter 1

Introduction

In this chapter, several issues concerning the design and implementation of controllers for flexible–link manipulators are discussed. The motivation and objectives of the monograph are given in Section 1.1. In Section 1.2, a literature review is provided. Section 1.3, discusses some experimental issues. In Section 1.4, an inverse dynamics control strategy is explained. Finally, in Section 1.5 an outline of the monograph is given.

1.1. Preamble

The problem of modeling and control of flexible–link manipulators has received much attention in the past several years. There are a number of potential advantages arising from the use of light–weight flexible–link manipulators. For instance, in designing a space manipulator, energy efficiency and microgravity must be considered. For this reason, the robot manipulator is normally designed as a light–weight structure which results in smaller actuators for driving the joints and consequently less energy is consumed. On the other hand, increased structural flexibility may be desirable in tasks such as cleaning a delicate surface or avoiding damage to the manipulator system due to accidental collisions [1, 2]. The use of light–weight manipulators also results in a high ratio of payload to arm weight [3]. Traditional manipulators have a poor load-carrying capacity – 5 to 10 percent of their own weight. This restriction is mainly imposed by the requirements for having a stable closed–loop system. By using heavy (rigid) robots, a designer makes the control problem less complicated. Light–weight manipulators also exhibit higher speed manipulations compared to conventional rigid manipulators.

Consequently, achieving high speed manipulation with lower energy consumption is a desirable objective. However, in taking advantage of the light–weight structure one should also be concerned with the inherent complexities involved. Structural flexibility causes extreme difficulty in modeling the manipulator dynamics and providing stable and efficient motion of the end-effector of the manipulator. This requires inclusion of deformation effects due to the flexibility of the arms in the dynamic equations which generally tends to complicate the analysis and design of the control laws. Flexible-link robot models belong to a class of distributed parameter systems described by partial differential equations (PDEs). The assumed modes and finite element methods are two common approaches currently used for approximating the PDEs by a system of ordinary differential equations (ODEs). A relatively large number of flexible modes are required to accurately model the system behavior in this case.

For a rigid manipulator, the tip trajectory is completely defined by the trajectory of the joint. Effective control of the joint is equivalent to good control of the tip. The situation is not as straightforward for a flexible manipulator and difficulties arise when one tries to track a specified end-effector position trajectory by applying the torque at the joint. In this case, the control difficulty is due to the non-colocated nature of the sensor and actuator positions which results in unstable zero dynamics [4, 5]. In other words, the nonlinear system is non–minimum phase. Therefore, the system has an unstable inverse. The non–minimum phase property of the flexible arm makes exact asymptotic tracking of a desired tip trajectory impossible, if one is to employ causal controllers [6]. Furthermore, model truncation, which arises due to a finite-dimensional representation of a distributed parameter system causes unmodeled dynamics to be present in the mathematical model of the system. Using a reduced–order model for the controller design may also lead to the phenomenon of control and observation spillover. Control spillover is the excitation of the residual modes by the control action, and observation spillover is the contamination of sensor readings by the residual modes. When both control and observation spillover are present, the closed–loop system may become unstable. Further complications arise because of the highly nonlinear nature of the system and the difficulty involved in accurately modeling various friction and backlash terms. Moreover, the coupling between the rigid dynamics and the flexible dynamics of the link may also cause stability problems. A change in the arm configuration and in the payload also leads to a change in the arm dynamics. This change has a serious degrading effect on the performance of the controllers.

The various difficulties involved in controlling a flexible–link manipulator can be summarized as follows:

- **Instability of the zero dynamics** related to the tip position which yields a non–minimum phase system.

- **Highly nonlinear** nature of the system.

- Requiring a **large number of states** for accurate modeling.

- **Unmodeled dynamics** due to model truncation and presence of various friction and backlash terms.

- **Variation of the payload.**

The non–minimum phase characteristic, coupling effects, nonlinearities, parameter variations and unmodeled dynamics all contribute to make the problem of controlling the end-effector of a flexible-link manipulator much more difficult than for a rigid-link manipulator. Control strategies that ignore these uncertainties and nonlinearities generally fail to provide satisfactory closed-loop performance.

The principal aim of this monograph is to address the above issues and to develop and *experimentally* evaluate strategies for controlling flexible–link manipulators in the presence of all of the aforementioned difficulties. The first objective is to develop our control schemes assuming some *a priori* nominal (inexact) dynamics. This is due to the fact that accurate modeling of flexible–link manipulators is extremely difficult. The next objective is to relax this assumption and design a control strategy that does not rely on any *a priori* knowledge about the system dynamics. A single flexible–link test–bed is considered for the experimental work. The hub of the manipulator exhibits a considerable amount of Coulomb friction and the link is relatively long (1.2 *m*) and is very flexible. These characteristics make control of this test–bed a challenging problem.

1.2.   Control Strategies

As we mentioned earlier, various complexities are involved in the control problem of light-weight manipulators. In an effort to reduce these complexities several researchers have proposed to perform local linearization of the equations of motion. A common approach in this case is a colocated proportional derivative (PD) control [7, 8]. By colocated, we mean the sensors and actuators are placed at the same location. It is shown in [9] that colocated PD control at either end of the arm guarantees stability of the system to parameter variations within a certain bound. By using this type of controller, however, the flexible modes of the system remain unaffected. Kotnik *et al.* [10] and Ge *et al.* [11] combined the joint PD controller with tip acceleration or strain measurements to damp out the vibrations of the flexible modes at the tip. In [12], the flexible beam was approximated by a spring–mass system, and based on that a PD–type controller was designed to achieve bounded–input bounded–output (BIBO) stable tip tracking performance.

Of the early experimental work in this area, the work of Canon and Schmitz [13] should be mentioned. They applied linear quadratic Gaussian (LQG) control by designing an optimal controller that assumes the availability of all the states of the system. Sakawa *et al.* [14] used linear quadratic (LQ) control to dampen the flexible modes while tracking the hub reference trajectory. Availability of all the states of the system is assumed. In the recent work of [9] an LQG/H_∞ controller was presented for a single flexible–link manipulator. While the flexible modes are damped out in the inner-loop by the LQG controller, the outer–loop H_∞ controller ensures stability of the system in the presence of uncertainties. Optimal control theory is also used in [15, 16, 17, 18] for controlling flexible-link manipulators.

Bayo [19] proposed a non-causal controller which acts before the tip starts moving and after the tip stops moving. Kwon and Book [20] proposed a decomposition of the inverse dynamics of the manipulator into a causal and an anti-causal system by using coordinate transformations. The causal part is integrated forward in time and the anti-causal part is integrated backward in time. These methods, however, require heavy computation and are limited to a linear approximation of the flexible-link manipulator. Nemir *et al.* [21] introduced the pseudo–link concept but have not addressed the non–minimum phase issue pertaining to the chosen output.

Input command shaping is also used for flexible–link robot control.

There are two shaping algorithms, the impulse shaping method and the command filtering. The impulse shaping algorithm was first introduced in [22, 23] and was later used by Hillsley and Yurkovich [24] to control the vibrations of flexible modes in large angle slewing maneuvers of a two-link flexible robot. This method essentially involves the convolution of a sequence of impulses with the reference input to suppress the vibration of flexible modes. The command filtering algorithm deals with filtering out the frequencies around the flexible modes. This method has been applied to vibration control of flexible-link manipulators by Magee and Book [25, 26], Tzes and Yurkovich [27]. Khorrami *et al.* [28] combined the rigid body based controller with input preshaping to control a two–link flexible manipulator. The validity of such methods depends on exact knowledge of the flexible structure dynamics. Such methods are open-loop strategies.

Although linear controllers may yield desirable closed-loop performance, their region of operation is limited due to the nonlinearities present in the original system. To have a wider region of operation, one has to take into account the nonlinearities. The most common approach to compensate for the nonlinear dynamics of a rigid manipulator is the so-called inverse dynamics or computed torque strategy. By employing this scheme, the manipulator dynamics is externally linearized and decoupled by the nonlinear controller introduced in the feedback loop. A servo controller is then constructed for the resulting decoupled linear model so that certain design specifications are satisfied. This scheme assumes exact cancellation of the manipulator dynamics by the nonlinear controller. Wang and Vidyasagar [4] have shown that the nonlinear flexible–link system is not in general input–state linearizable. However, the system is locally input-output linearizable but the associated zero dynamics are unstable when the tip position is considered as an output of the system, that is, the system is non–minimum phase. Consequently, the extension of this approach to flexible-link manipulators is impeded by the non–minimum phase characteristic of the arm. There have been several methods proposed in the literature to resolve this issue.

For instance, singular perturbation theory [29, 30] has been used by several researchers for modeling and control of flexible-link manipulator. This method is attractive because of the two time-scale nature of the system dynamics. In these control strategies, a linear stabilizer is used to stabilize the fast dynamics (flexible modes) and a nonlinear controller is used to make the slow dynamics (joint variables) track the desired trajectories. In [31], a singular perturbation model for the case of multi–link manipulators was introduced which follows a similar approach in terms of modeling as that introduced by Khorasani and Spong [32] for the case of flexible–joint manipulators. The singular perturbation approach is also considered in [33, 34, 35, 36, 37]. A comparison is made experimentally between some of these methods by Aoustin *et al.* [38]. Standard singular perturbation results that are applied to flexible-link manipulators generally exclude high–performance light–weight manipulators, since a reduced–order rigid body equivalence of the flexible arm has limited use and application. To overcome this limitation, several researchers used the integral manifold approach introduced in [32, 39] to control the flexible-link manipulator [40, 41, 42]. In [40], a linear model of the single flexible–link manipulator was considered. A non–linear model of a two-link flexible manipulator is used in [41]. In this approach, new fast and slow outputs are defined and the original tracking problem is reduced to track the slow output and stabilize the fast dynamics.

The output redefinition approach is also used to overcome the non–minimum phase characteristic of the system. A number of approaches have been proposed in [43] for this purpose. In [44], the output of the system was redefined as the "reflected" tip position to ensure stable zero dynamics for the new input–output map. This method was later used in [45, 46] to control a single flexible–link manipulator using the passivity theorem. De Luca and Lanari [47] studied the regions of sensor and actuator locations for achieving the minimum phase property for a single flexible link. Similar approaches were used in [48, 49] by selecting a point on the link between the joint and the tip as a new output. The new output is defined in such a way that the zero dynamics associated with this output are stable. Based on the new output, the input–output linearization (inverse dynamics) approach [50, 51] was used to control a single–link manipulator [48] and a two–link manipulator [49, 52]. In [53, 54, 55], a novel approach based on transmission zero assignment [56, 57] was applied to control a single flexible–link manipulator. The idea is to add a feedthrough compensator in parallel to the plant so that the zeros of the new output are placed at specified locations in the left–half plane. Based on this output, a dynamic output feedback strategy can be used to place all the poles of the closed-loop system at the desired locations. Using the output redefinition approach to control some classes of nonlinear non–minimum phase system was also suggested in [58, 59]. In [60, 61], the authors applied this method to a flight control system.

It should be pointed out that all of these methods assume exact knowledge of the dynamics and the nonlinearities of the flexible-link system. Since, in general, it is very difficult to model a flexible-link manipulator accurately, the performance of these control strategies may be unsatisfactory for real applications.

In [62, 63, 64, 65], the authors studied fuzzy supervisory controllers that tune conventional (PID) controllers. In [66], a fuzzy logic controller for a fast-moving single flexible-link was developed which focused on smooth, rigid body motion control. In [67], a fuzzy learning control approach was used focusing on automatic synthesis of a direct fuzzy controller and its subsequent tuning when there are payload variations. In [68, 69], a fuzzy logic supervisory level is used for a lower-level conventional controller selection. In [70], a two-level hierarchical rule-based controller was proposed. This scheme employs an upper-level "expert controller" that captures the knowledge about how to supervise the application of low-level fuzzy controllers during movements in the robot workspace. In [71], an "anticipatory fuzzy control" for a flexible beam was introduced. By "anticipatory", we mean that the effect of the control signal on the system output is predicted and is replaced with another controller if it is not acceptable. This prediction is accomplished through mathematical simulation of the dynamic equations. A neural network is trained to perform the function of the anticipatory fuzzy logic controller.

An approach that looks promising for control of flexible-link manipulators is based on neural networks. There are several ways that a neural network-based controller can be used to adaptively control rigid manipulators. Most of them, however, are based on either the minimum phase characteristic of the input-output map or require full state measurements- conditions that are not generally satisfied for flexible-link manipulators. In [72] a neural network controller for a flexible-link manipulator was designed. Hub position and velocity were used to stabilize the system. Then an adaptive observer was used to identify the system parameters. A modified Hopfield neural network was used to realize the observer. Based on the identified parameters, an auxiliary system was constructed and a feed-forward controller was designed so that the output of the auxiliary system tracks the desired trajectory. This work, however, is restricted only to linear models of the flexible-link robot. The problem of online self tuning adaptive control (OLSTAC) using a back propagation neural network (BPNN) is considered in [73]. The authors employed online learning of BPNN in both stages of OL-STAC, i.e., system identification as well as control. It was assumed that the dynamics of the flexible–link manipulator can be separated into two nonlinear/uncertain terms representing the state term and the control term. The control consists of two parts, the displacement part, and the velocity part. Two separate neural networks were employed to construct these signals. The error function was obtained by propagating back the error between the desired trajectory and the output of the neural identifier, through the neural identifier. In [74], partial knowledge of the dynamics of the flexible-link is assumed and the unknown part of the dynamics is identified by a supervised learning algorithm. The same methodology used in [75] was pursued for control. The control is constructed in two stages, an optimal controller and an unsupervised neural network controller using model-based predictive control. The scheme is based on an identification stage that also requires feedback from the states of the system. In [76], a variable structure control was used. A neural network was employed to identify the payload and to select the proper linear controller previously designed for each range of the payload variation. Register *et al.* [77] extended the neural controller proposed in [78] to a lightly damped system by adding a term related to the hub velocity to the cost function of the neural network. In [79], the joint tracking control problem for a space manipulator using feedback–error learning is proposed. They assume known rigid body nonlinearities and neglected flexible dynamics. The known nonlinearities were used as composite inputs to the neural network. In this scheme, tip position tracking is not guaranteed especially for high–speed desired trajectories. In [80], tracking control of a partially known flexible-link robot was considered. A physically meaningful and measurable output, namely hub angle, was selected for tracking. The controller was composed of a singular-perturbation-based fast control and an outer-loop slow control. The slow subsystem is controlled by a neural network for feedback linearization, plus a PD outer-loop for tracking and a robustifying term to ensure the closed-loop stability. This work is restricted to a manipulator with sufficiently large stiffness. It should be mentioned that except for the work in [79] which considered the joint tracking problem assuming known rigid–body nonlinearities, no experimental work has been reported for controlling flexible-link manipulators using neural networks.

1.3.   Experimental Issues

In this section, we briefly discuss the complexities inherent in practical implementation of controllers for flexible-link manipulators. These complexities may basically be categorized as follows:

- Hub friction and stiction: Because of the lightweight structure, the joint torque is lower than in industrial manipulators and thus joint friction is relatively higher. This can cause difficulty in modeling and result in poor performance for model-based control strategies.

- Vibration of the flexible modes: As the speed of the desired trajectory is increased, the magnitude of the vibration of the flexible modes becomes larger and this may result in instability of the system due to the limitation of the sensors usually used for measuring these vibrations. This is important especially for linear controllers that use higher gains to satisfy the design specifications.

- Computation time required for implementing the algorithms and sampling rate selection: Discretization of analog controllers usually demands high servo rates which, due to the complexity of the nonlinear controllers, may not always be feasible.

- Actuator saturation: Different control strategies demand different amounts of input torques. If the actuator cannot provide the demanded torque, it may lead to instability of the closed-loop system.

- Type of sensors used for deflection measurements: Another major question in implementing a controller is the number and type of sensing points. Since tip positions and their time derivatives are to be controlled, the least information to be provided to the controller is accurate information on the tip positions. In particular. Sensing the tip deflection is possible by using strain gauges and/or photodetectors. Accuracy, installation, dynamic range and the possibility of measuring all the states are important issues to be considered. For instance, using photodetectors may result in more accurate sensing; however, feeding the state vector back for control purposes will not be possible by using only photodetectors.

On the other hand, using strain gauges brings certain difficulties that may affect the accuracy of the measurement. For instance, one needs to determine the mode shapes for the computation of the flexible modes from the measurements.

1.4. Inverse Dynamics

Consider the following single–input, single–output (SISO) affine nonlinear system [1] described by the state space representation

$$\begin{cases} \dot{\mathbf{x}} = \mathbf{f}(\mathbf{x}) + \mathbf{g}(\mathbf{x})u \\ y = h(\mathbf{x}) \end{cases} \tag{1.1}$$

where $\mathbf{x} \in \Re^n$, $y \in \Re$, $u \in \Re$ and $\mathbf{f}, \mathbf{g}, h$ are smooth nonlinear functions. The objective is to make the output y track a desired trajectory y_r while keeping the whole state bounded, where y_r and its time derivatives up to a sufficiently high order are assumed to be known and bounded. In input–output linearization (inverse dynamics) approach, this goal is achieved by generating a linear differential relation between the output y and a new input ν. The new input ν is later designed to satisfy tracking requirements.

Relative Degree

An apparent difficulty in controlling system (1.1) is that the output y is only indirectly related to the input u, through the state variable $\mathbf{x}$ and nonlinear equations (1.1). Therefore, it is difficult to see how the input u can be designed to control the tracking behavior of the output y. If a direct and simple relation can be found between the system output y and the control input u, then the difficulty of the tracking control design can be reduced. This idea constitutes the basis for the so–called input–output linearization approach to nonlinear control design.

To generate a direct relationship between the output y and the input u, the output function y is repeatedly differentiated until the input u appears. Suppose the input u appears after the rth differentiation, then we have

$$y^{(r)} = a(\mathbf{x}) + b(\mathbf{x})u.$$

The system is said to have *relative degree* r in a region of interest $\Omega \in \Re^n$ if $\forall \mathbf{x} \in \Omega$, $b(\mathbf{x}) \neq 0$.

---

[1] A class of nonlinear systems that are linear in control.

Normal Forms

Assuming the relative degree r is defined and $r < n$, then by using $y, \dot{y}, \cdots y^{r-1}$ as part of the new state component

$$\mu = \begin{bmatrix} \mu_1 & \mu_2 \cdots & \mu_r \end{bmatrix}^T = \begin{bmatrix} y & \dot{y} \cdots y^{(r-1)} \end{bmatrix}^T$$

and by defining a (local) diffeomorphism [6] for a state transformation

$$\begin{bmatrix} \mu_1 & \cdots & \mu_r & \psi_1 & \cdots & \psi_{n-r} \end{bmatrix}^T = \mathbf{\Pi}(\mathbf{x}),$$

the *normal form* of the system can be written as

$$\begin{cases} \dot{\mu}_1 &= \mu_2 \\ &\vdots \\ \dot{\mu}_{r-1} &= \mu_r \\ \dot{\mu}_r &= a(\mu, \mathbf{\Psi}) + b(\mu, \mathbf{\Psi})u \end{cases} \tag{1.2}$$

$$\dot{\mathbf{\Psi}} = \mathbf{Z}(\mu, \mathbf{\Psi}) \tag{1.3}$$

$$y = \mu_1$$

where μ and $\mathbf{\Psi} = \begin{bmatrix} \psi_1 & \psi_2 \cdots & \psi_{n-r} \end{bmatrix}^T$ are the *normal coordinates* and a, b and $\mathbf{Z}$ are nonlinear functions of μ and $\mathbf{\Psi}$.

The Zero Dynamics

By employing the inverse dynamics approach, the dynamics of a non-linear system are decomposed into an external (input–output) part and an internal ("unobservable") part. System (1.2) is called the external dynamics and system (1.3) is called the internal dynamics. Since the external part consists of the controllability canonical form between y and u, as in (1.2), it is easy to design the input u to cancel the non-linearities $a(\mu, \mathbf{\Psi})$ and $b(\mu, \mathbf{\Psi})$ or equivalently to externally linearize (1.2). Towards this end, the input u should be taken as

$$u = \frac{1}{b(\mu, \mathbf{\Psi})}[-a(\mu, \mathbf{\Psi}) + \nu] \tag{1.4}$$

where ν is the control for the linearized system to be designed subsequently. Specifically, to make the system output y track the desired trajectory y_r asymptotically, the servo control ν is designed as

$$\nu = y_r^{(r)} - k_{r-1}(\mu_{r-1} - y_r^{(r-1)}) - \cdots - k_0(\mu_1 - y_r) \tag{1.5}$$

where $k_0, \cdots, k_{r-1}$, are selected to place the roots of $s^r + k_{r-1}s^{r-1} + \cdots + k_1 s + k_0 = 0$ at some desired location in the left half of the complex plane (LHP). Although the nonlinear control law (1.4) and the servo control law (1.5) seem to be sufficient to guarantee the desired trajectory performance, the question is whether the internal dynamics will behave well, *i.e.*, whether the internal states will remain bounded. Since the control design must account for the whole dynamics (and therefore cannot tolerate the instability of internal dynamics), the internal behavior has to be addressed carefully.

Generally, the internal dynamics depend on the state μ. However, some conclusions about the stability of the internal dynamics can be made by studying the so-called *zero dynamics*. The zero dynamics (1.6) are defined as the internal dynamics when the control input u is such that the output y is maintained at zero. In order for the system to operate in zero dynamics, the initial state of the system must be selected as $\mu = 0$ and $\Psi(0) = \Psi_0$ (arbitrary). Furthermore the input u must be such that y stays at zero, which implies that $y^{(r)}(t) = 0$. From (1.2), this means that u must equal

$$u = \frac{-a(0, \Psi)}{b(0, \Psi)}$$

Now, the system dynamics can be simply written in the normal form as

$$\begin{aligned} \dot{\mu} &= 0 \\ \dot{\Psi} &= Z(0, \Psi). \end{aligned} \tag{1.6}$$

Recall that linear systems whose zero dynamics are stable are called minimum phase. This notion can be extended to nonlinear systems. The nonlinear system (1.1) is said to be *asymptotically minimum phase* if its zero dynamics (1.6) are asymptotically stable. Then the following result from [6] shows that provided the *zero dynamics are asymptotically stable*, the control law (1.4) and (1.5) *locally* stabilizes the whole system.

Theorem 1.1 *[6] Assume that the system (1.1) has relative degree r and that its zero dynamics are locally asymptotically stable. Choose constants k_i such that the polynomial $K(s) = s^r + k_{r-1}s^{r-1} + \cdots + k_1 s + k_0$ has all its roots strictly in the LHP. Then, the control laws (1.4) and (1.5) yield a locally asymptotically stable closed–loop system.*
Proof: *For details refer to [6].*

Therefore, instability of the zero dynamics may imply instability of the closed–loop system. Hence, the inverse dynamics approach cannot be applied directly to non–minimum phase systems.

For a linear time–invariant system, b is a constant matrix, a is a linear function of the states and the internal dynamics and consequently the zero dynamics are specified by a linear time–invariant system. Note that in this case, the zero dynamics are exclusively determined by the location of the system transmission zeros. Therefore, inverse dynamics control is directly applicable only if the system zeros are all in the LHP.

1.5. Outline

Chapter 2: Manipulator Model

In this chapter, the dynamic equations of the manipulator are derived. Although, the proposed control strategies are not based on a full non-linear model, derivation of the dynamic equations is still needed for the purpose of simulations (forward dynamics). The model is derived by using the Recursive Lagrangian approach and then verified by using experimental data. The first two control schemes proposed in this monograph assume *a priori* knowledge about the linear model of the manipulator. Verification of this linear model is difficult due to the presence of significant amount of stiction and Coulomb friction at the hub. To resolve this problem, experimental verification of the linear model is performed by using PD hub position control with high gain to overcome the effect of the friction. At the end of this chapter, the non-minimum phase property of the flexible–link manipulator is discussed.

Chapter 3: Output Redefinition

The output redefinition approach is illustrated in this chapter. The idea of output redefinition is motivated by the fact that zero dynamics depend on the choice of the output. More specifically, the output is redefined such that the linearized zero dynamics associated with the new output are asymptotically stable. First, the "reflected tip position" introduced in [44] is discussed. It is shown that the difference between the reflected tip position and the actual tip position becomes significant for high–speed reference trajectories as well as for very flexible–link manipulators. Next, the sum of the joint angle and a scaling of the tip elastic deformation is defined as a new output and the condition for obtaining this output is provided. The new output is defined assuming a minimum *a priori* knowledge about the system dynamics. Specifically, no *a priori* knowledge about the payload mass is assumed. This forms the basis for the construction of the first two control strategies proposed in the next chapter.

Chapter 4: Proposed Neural Network Structures

This chapter presents the proposed neural network–based control schemes. Four neural network schemes that utilize the output redefinition approach are introduced. The first two schemes are developed by using a modified version of the "feedback–error learning" approach to learn the inverse dynamics of the flexible manipulator. Both schemes assume some *a priori* knowledge of the linear model of the system. This assumption is relaxed in the third and fourth schemes. In the third scheme, the controller is based on tracking the hub position while controlling the elastic deflection at the tip. The fourth scheme employs two neural networks, one of the neural networks define an appropriate output for feedback and the other neural network acts as an inverse dynamics controller. Simulation results for two single flexible–link manipulators and a two–link manipulator are presented. Finally, the last two schemes are modified to improve the transient and steady-state responses of the system, and to increase the dynamic range of the controller and the robustness of the system to changes in both magnitude and frequency of the desired trajectory.

Chapter 5: Experimental Results

Experimental results are demonstrated in this chapter. First, an experimental test–bed is described which consists of a very flexible link. The actuating and sensing mechanisms; controller, data acquisition and interface card are discussed. Different aspects of implementation of the proposed control strategies and inherent difficulties involved are discussed. Experimental results are provided that are performed in the presence of payload variations, friction, and model uncertainties.

Chapter 2

Manipulator Model

2.1. Introduction

This chapter develops the dynamic model for a flexible–link manipulator. Although the proposed control strategies in this monograph require no knowledge or *only* a partial knowledge about the system dynamics, the analytical model of the system is still needed for the purpose of simulations (forward dynamics). Based on Euler–Bernoulli beam theory, a partial differential equation (PDE) known as the Euler–Bernoulli beam equation is used to model the vibration of the beam, yielding an expression for the deflection as a function of time and distance along the beam. Using the separation of variables method, the beam equation is expressed as two ordinary differential equations (ODEs). The Recursive Lagrangian approach is then used to derive the dynamic equations of the manipulator. This model is then linearized and is used in the first two proposed schemes in later chapters.

The dynamic modeling of the manipulator is derived in Section 2.2. In Section 2.3, the non–minimum phase property of the flexible–link manipulator is discussed. Finally, in Section 2.4, the experimental manipulator's parameters are incorporated into the model. The accuracy of the model is verified by comparing its responses with those of the experimental manipulator.

2.2. Dynamic Modeling

Systematic methods for formulating the dynamic equations of motion for rigid manipulator arms have been extensively studied by many re-

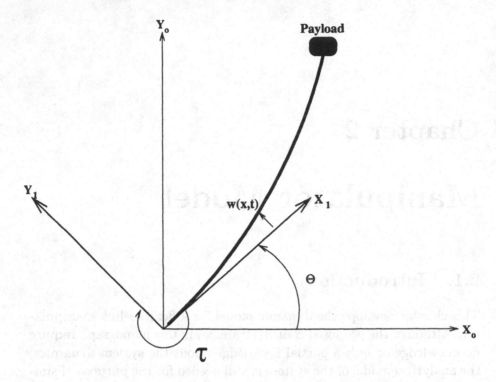

Figure 2.1. Schematic of the one link flexible arm.

searchers and variety of techniques are available to obtain the dynamic equations e.g. [81, 82]. Modeling of a flexible-link manipulator, however, is difficult due to the structural flexibility and the distributed parameter nature of the system. The assumed modes and finite element methods are two common ways to obtain an approximate model of a flexible–link manipulator.

In the following, we focus on the modeling of a single flexible–link manipulator which is fixed at one end (hub) and is driven by a torque τ. The other end is free to flex in a horizontal plane, and has a mass M_l as a payload (see Figure 2.1). It is assumed that the length of the beam, l, is much greater than its width, thus restricting the beam to oscillate in the horizontal direction. Neglecting the effects of shear deformation and rotary inertia , the deflection of any point on the beam $W(x,t)$ is given by the Euler-Bernoulli beam equation (e.g. [83])

$$EI\frac{\partial^4 W(x,t)}{\partial x^4} + \rho A l^4 \frac{\partial^2 W(x,t)}{\partial t^2} = 0 \qquad (2.1)$$

where E is Young's modulus of the material, A is the link cross–sectional area, I is its inertia and ρ is its uniform density.

Before a dynamic model of the arm can be developed, it is necessary to find the natural modes of the arm. The natural modes vary with the rigid dynamics and the configuration of the arm and, therefore, require frequent recalculation [3]. It is possible, however, to model the system using a set of modes other than natural modes. As long as the geometric boundary conditions remain unchanged, the same set of functions can be applied to the arm throughout its workspace. Two commonly used sets of functions are the so–called cantilever (clamped–free) modes [84, 85, 86, 44] and pinned–free modes [84, 13, 85].

Hasting and Book [86] experimentally verified that clamped–mass admissible functions yield better results than other mode shapes such as pinned-free modes used by Canon and Schmitz [13]. Bellezza et al. [84] have shown that open–loop modes obtained by using clamped–free and pinned–free eigenfunctions are identical and only differ in the reference frame in which the elastic deflections are measured. Cetinkunt and Yu [85] have compared the first three modes of the closed–loop system for pinned–free and clamped–free mode shapes with the modes obtained from the exact solution of the Euler–Bernoulli beam equation. They have shown that the predictions of clamped–free mode shapes are much more accurate than the predictions of pinned–free mode shapes.

To obtain a solution for equation (2.1), separation of variables is used, i.e.

$$W(x, t) = \phi(x)\delta(t) \qquad (2.2)$$

the general solution for equation (2.1) can be written as the following [87, 88]:

$$
\begin{aligned}
\delta(t) &= Ae^{j\omega t} \\
\omega^2 &= \frac{\beta^4 EI}{\rho Al^4} \\
\phi(x) &= C_1 \sin(\beta x) + C_2 \cos(\beta x) + C_3 \sinh(\beta x) + C_4 \cosh(\beta x)
\end{aligned}
\qquad (2.3)
$$

A complete solution of the cantilever beam problem with an inertia tip load consisting of point mass M_l requires solving for the constants C_1 through C_4 in equation (2.3). These constants are calculated using the following boundary conditions:
At $x = 0$

$$W(x, t)\,|_{x=0} = 0$$

$$\frac{\partial W(x,t)}{\partial x}\Big|_{x=0} = 0$$

at $x = l$

$$(\text{Bending moment}) M = EI\frac{\partial^2 W(x,t)}{\partial x^2}\Big|_{x=l} = -J_l\frac{\partial^2}{\partial t^2}\left(\frac{\partial W(x,t)}{\partial x}\right)\Big|_{x=l}$$

$$(\text{Shear force}) S = EI\frac{\partial^3 W(x,t)}{\partial x^3}\Big|_{x=l} = M_l\frac{\partial^2 W(x,t)}{\partial t^2}\Big|_{x=l} \quad (2.4)$$

where J_l is the load inertia. The conditions on M and S follows from the fact that M_l and J_l cause concentrated force $= M_l\frac{\partial^2 W(x,t)}{\partial t^2}$ and bending moment to act on the beam at $x = l$.

Using (2.2), (2.3) and (2.4), boundary conditions at the end–point become:

$$\frac{\partial^2 W(x,t)}{\partial x^2}\Big|_{x=l} = \frac{J_l\beta^4}{\rho Al^3}\frac{\partial W(x,t)}{\partial x}\Big|_{x=l}$$

$$\frac{\partial^3 W(x,t)}{\partial x^3}\Big|_{x=l} = -\frac{M_l\beta^4}{\rho Al}W(x,t)\Big|_{x=l}$$

The clamped conditions at the joint yield

$$C_3 = -C_1 \qquad\qquad C_4 = -C_2$$

while the mass conditions at the end–point lead to

$$\begin{bmatrix} \mathcal{Q}(\beta) \end{bmatrix}\begin{bmatrix} C_1 \\ C_2 \end{bmatrix} = 0$$

The frequency equations are then given by setting to zero the determinant of the 2×2 matrix $\mathcal{Q}(\beta)$. Now, the positive values of β are obtained by the solutions of the transcendental equation

$$(\ 1 + cos\beta\cosh\beta) - M\beta(sin\beta\cosh\beta - cos\beta sinh\beta)$$
$$- J\beta^3(sin\beta\cosh\beta + cos\beta sinh\beta) + MJ\beta^4(1 - cos\beta\cosh\beta) = 0$$

where $M = M_l/\rho Al$ and $J = J_l/\rho Al^3$. For $M_l = J_l = 0$, the above equation is reduced to

$$1 + \cos\beta\cosh\beta = 0.$$

The natural frequencies of vibration are obtained from

$$\omega_i = \beta^2\sqrt{\frac{EI}{\rho Al^4}}.$$

To each of the natural frequencies ω_i, corresponds a specific mode shape function $\phi_i(x)$ and a specific amplitude function or *normal coordinate* $\delta_i(t)$. A single solution to the deflection problem may be expressed as

$$w_i(x,t) = \phi_i(x)\delta_i(t).$$

The single solution $w_i(x,t)$ will not usually satisfy the initial conditions for position, $W(x,0)$ and velocity $\dot{W}(x,0)$. Since equation (2.1) is linear and homogeneous, the principle of superposition holds that the sum of infinitely many solutions w_i is a solution of (2.1).

$$W(x,t) = \sum_{i=1}^{\infty} \phi_i(x)\delta_i(t). \tag{2.5}$$

The set of eigenfunctions $\{\phi_i(x),\ i = 1, 2, \cdots, \infty\}$ is the set of functions

$$\phi_i(x) = c_i[(sin\beta_i x - sinh\beta_i x) - \frac{(sin\beta_i l + sinh\beta_i l)}{(cos\beta_i l + cosh\beta_i l)}(cos\beta_i x - cosh\beta_i x)].$$

The constant c_i normalizes the eigenfunctions so that

$$\int_0^l \phi_i(x)^2 dx = 1.$$

2.2.1. The Assumed Modes Method

Equation (2.5) states that an exact solution to the Euler–Bernoulli PDE requires an infinite number of modes. The exact solution, however, can be approximated by the assumed modes method using a finite number of modes,

$$W(x,t) = \sum_{i=1}^{n} \varphi_i(x)\delta_i(t)$$

where the $\varphi_i(x)$ are any functions that satisfy the boundary conditions. If the functions $\varphi_i(x)$ are selected to be a set of polynomials in x, the resulting frequencies of vibration are only approximations to the actual natural frequencies of the system. If the functions $\varphi_i(x)$ are chosen as the eigenfunctions $\phi_i(x)$, the frequencies of vibration equal the natural frequencies of the system. Therefore, the deflection of the cantilever beam can be approximated by

$$W(x,t) = \sum_{i=1}^{n} \phi_i(x)\delta_i(t).$$

2.2.2.   Dynamic Equations

By considering a finite number n of modal terms, the dynamic equations for the one-link flexible arm can be derived by using a Recursive Lagrangian approach [89]

$$M(\delta) \begin{bmatrix} \ddot{\theta} \\ \ddot{\delta} \end{bmatrix} + \begin{bmatrix} h_1(\dot{\theta}, \delta, \dot{\delta}) + F_1 \dot{\theta} + f_c \\ h_2(\dot{\theta}, \delta) + K\delta + F_2 \dot{\delta} \end{bmatrix} = \begin{bmatrix} u \\ 0 \end{bmatrix}, \qquad (2.6)$$

where θ is the hub angle, δ is the $n \times 1$ vector of deflection variables, and h_1 and h_2 are the terms representing the Coriolis and the centrifugal forces, namely

$$\begin{aligned} h_1(\dot{\theta}, \delta, \dot{\delta}) &= 2M_l \dot{\theta}(\Phi_e^T \delta)(\Phi_e^T \dot{\delta}) \\ h_2(\dot{\theta}, \delta) &= -M_l \dot{\theta}^2(\Phi_e \Phi_e^T)\delta \end{aligned} \qquad (2.7)$$

M is the positive–definite symmetric inertia matrix given by

$$M(\delta) = \begin{bmatrix} m_{11}(\delta) & M_{12}^T \\ M_{12} & M_{22} \end{bmatrix}.$$

The elements m_{ij} of the inertia matrix $M(\delta)$ take on the expression below, which is valid when clamped–free modes are used in the presence of a tip payload and neglecting load inertia.

$$\begin{aligned} m_{11}(\delta) &= J_0 + M_l l^2 + I_h + M_l(\Phi_e^T \delta)^2 & (2.8) \\ m_{1j} &= M_l l \phi_{j-1,e} + \sigma_{j-1}, \quad j = 2, \cdots, n+1 \\ m_{ii} &= \rho A + M_l \phi_{i-1,e}^2, \quad i = 2, \cdots, n+1 \\ m_{ij} &= M_l \phi_{i-1,e} \phi_{j-1,e}, \quad i = 2, \cdots, n+1, \quad j \neq i \end{aligned}$$

with

$$\begin{aligned} \Phi_e^T &= \Phi^T|_{x=l} = [\phi_1 \cdots \phi_n]|_{x=l} \quad \phi_{ie} = \phi_i(x)|_{x=l} \\ \sigma_i &= \rho A \int_0^l \phi_i(x) x \, dx \end{aligned}$$

where m is the link mass, I_h is the joint actuator inertia, and J_0 is the link inertia relative to the joint; K is the positive-definite diagonal stiffness matrix

$$\begin{aligned} K &= diag\{k_1, \cdots, k_n\} \\ k_i &= EI \int_0^l [\frac{\partial^2 \phi_i(x)}{\partial x^2}]^2 dx \end{aligned} \qquad (2.9)$$

F_1 is the viscous damping at the hub and F_2 is the positive-definite diagonal damping matrix

$$F_2 = diag\{f_1, \cdots, f_n\} \tag{2.10}$$

which accounts for the internal viscous friction in the flexible structure.

As can be seen, the hub friction f_c is included in equation (2.6). The hub friction cannot be modeled accurately and is included for simulation purposes only. In this regard, f_c is often considered as the Coulomb friction which may be represented by a hard nonlinearity $f_c = C_{coul}SGN(\dot{\theta})$, where

$$SGN(\dot{\theta}) = \begin{cases} 1 & \text{for } \dot{\theta} > 0, \\ -1 & \text{for } \dot{\theta} < 0, \\ 0 & \text{for } \dot{\theta} = 0 \end{cases} \tag{2.11}$$

or may be approximated by a smooth function and is expressed as

$$f_c = C_{coul}(\frac{2}{1 + e^{-a\dot{\theta}}} - 1), \quad C_{coul} > 0, \quad a > 0.$$

The input torque τ is represented with u.

The Recursive Lagrangian approach can also be used to derive the dynamic equations for multi-link flexible manipulators as

$$M(\theta, \delta) \begin{bmatrix} \ddot{\theta} \\ \ddot{\delta} \end{bmatrix} + \begin{bmatrix} f_1(\theta, \dot{\theta}) + h_1(\theta, \dot{\theta}, \delta, \dot{\delta}) + F_1\dot{\theta} + f_c \\ f_2(\theta, \dot{\theta}) + h_2(\theta, \dot{\theta}, \delta, \dot{\delta}) + K\delta + F_2\dot{\delta} \end{bmatrix} = \begin{bmatrix} u \\ 0 \end{bmatrix} \tag{2.12}$$

where θ is the $n \times 1$ vector of joint variables, δ is the $m \times 1$ vector of deflection variables and f_1, f_2, h_1 and h_2 are the terms due to gravity (f_1 only), Coriolis, and centrifugal forces. If a finite number of flexible modes m_i is considered for ith link, then $\delta = \begin{bmatrix} \delta_1 & ... & \delta_n \end{bmatrix}^T$, $\delta_i = \begin{bmatrix} \delta_{i1} & ... & \delta_{im_i} \end{bmatrix}, i = 1, \cdots, n$, and $m = \sum_{i=1}^{n} m_i$.

As pointed out in [44], the tip position can be obtained from $y_{ti} = \theta_i + \frac{W_i(l_i, t)}{l_i}$, where $W_i(l_i, t)$ is the elastic deflection at the tip and l_i is the length of the ith link. When m_i modes are considered for the ith link, $W_i(l_i, t)$ may be expressed as $W_i(l_i, t) = \sum_{j=1}^{m_i} \phi_{ij}(l_i)\delta_{ij}(t)$, where ϕ_{ij} is the jth eigenfunction of the ith link and δ_{ij} is the jth mode of the ith link. Thus, the tip position vector can be expressed as

$$y_t = \theta + \gamma_{n \times m}\delta$$

where

$$\gamma_{n \times m} = \begin{bmatrix} v_1^T & 0 & \cdots & 0 \\ 0 & v_2^T & \cdots & 0 \\ \vdots & \vdots & \vdots & \vdots \\ 0 & 0 & \cdots & v_n^T \end{bmatrix}, \tag{2.13}$$

$$v_i^T = \frac{1}{l_i} \left[\phi_{i1}(l_i) \cdots \phi_{imi}(l_i) \right], i = 1, \cdots, n$$

$$y_t^T = \left[y_1 \cdots y_n \right]$$

2.2.3.   Local Linearization of the Equations of Motion

Local linearization may be used to derive a linear model of the actual system. This linear model is an approximation of the nonlinear system in the neighborhood of an operating point. Neglecting the Coulomb friction term f_c in (2.6) and performing local linearization yields

$$M_0 \begin{bmatrix} \ddot{\theta} \\ \ddot{\delta} \end{bmatrix} + \begin{bmatrix} F_1 \dot{\theta} \\ K\delta + F_2 \dot{\delta} \end{bmatrix} = \begin{bmatrix} u \\ 0 \end{bmatrix} \tag{2.14}$$

where M_0 is the linearized M. Now, by defining $\mathbf{X^T} = [x_1, \cdots, x_{2n+2}] = [\theta, \delta^T, \dot{\theta}, \dot{\delta}^T]$, the state space equations can now be derived in the form

$$\dot{X} = AX + Bu$$
$$y_t = C_t X \tag{2.15}$$

where

$$A = \begin{bmatrix} 0_{n+1 \times n+1} & I_{n+1 \times n+1} \\ -M^{-1}KK & -M^{-1}FF \end{bmatrix},$$

$$B = \begin{bmatrix} 0_{n+1 \times 1} \\ M^{-1} \begin{bmatrix} u \\ 0_{n \times 1} \end{bmatrix} \end{bmatrix},$$

$$C_t = [1 \; \frac{\phi_1}{l} \; \cdots \; \frac{\phi_n}{l} \; 0 \; \cdots \; 0],$$

$$KK = \begin{bmatrix} 0 & 0_{1 \times n} \\ 0_{n \times 1} & K \end{bmatrix},$$

$$FF = \begin{bmatrix} F_1 & 0_{1 \times n} \\ 0_{n \times 1} & F_2 \end{bmatrix}.$$

Wang and Vidyasagar [44] have shown that if the number of flexible modes is increased the transfer function from the input torque to the tip position does not have a well defined relative degree. Moreover, the associated zero dynamics are unstable for this output. Hence the system is non–minimum phase and is very difficult to control using this output for feedback. This issue is discussed in the following section.

2.3. Non–minimum Phase Characteristic

The non–minimum phase behavior is a characteristic of underactuated mechanical systems such as the *acrobot* [90], PVTOL aircraft [61, 6], and the flexible–link manipulator. In the flexible–link manipulator, the non–minimum phase behavior may be traced to the non–colocated nature of the sensor and actuator position.

To illustrate the non–minimum phase characteristic of the flexible–link system, a single–link flexible arm is considered (see Figure 2.1). Equation (2.14) represents a linearized model of the system. By setting $M_l = 0$, and by considering one flexible mode, Equation (2.14) can be rewritten as

$$m_{11}\ddot{\theta} + m_{12}\ddot{\delta} + F_1\dot{\theta} = u$$
$$m_{12}\ddot{\theta} + m_{22}\ddot{\delta} + F_2\dot{\delta} + K\delta = 0 \qquad (2.16)$$

where m_{ij}, $i, j \in \{1, 2\}$ are elements of the inertia matrix M_0 (linearized version of (2.8)), F_1 is the viscous damping at the hub and K and F_2 are given by (2.9) and (2.10), respectively.

$$m_{11} = J_0 + I_h,$$
$$m_{12} = \rho A \int_0^l \phi_1(x)x\,dx,$$
$$m_{22} = \rho A$$
$$K = EI \int_0^l [\frac{\partial^2 \phi_1(x)}{\partial x^2}]^2 dx \qquad (2.17)$$

Defining the output as the tip–position,

$$y_t = \theta + \frac{\phi_{1e}}{l}\delta,$$

the zero dynamics related to this output can be found by the procedure explained in Section 1.4. Setting y_t identically zero yields

$$\theta = -\frac{\phi_{1e}}{l}\delta. \qquad (2.18)$$

Substituting (2.18) in (2.16) gives

$$(m_{12} - \frac{\phi_{1e}}{l}m_{11})\ddot{\delta} + F_1\dot{\theta} \;\; = \;\; u \tag{2.19}$$

$$(m_{22} - \frac{\phi_{1e}}{l}m_{12})\ddot{\delta} + F_2\dot{\delta} + K\delta \;\; = \;\; 0 \tag{2.20}$$

Now, the stability of the zero dynamics (2.20) can be investigated by specifying the sign of the coefficient $(C_{z0} = m_{22} - \frac{\phi_{1e}}{l}m_{12})$. Using (2.17), C_{z0} can be written as

$$\rho A(1 - \frac{\phi_{1e}}{l}\int_0^l x\phi_1(x)dx). \tag{2.21}$$

Since the product of $x\phi_1(x)$ for clamped–free eigenfunction is always positive, if l is large enough C_{z0} becomes negative in (2.21). Therefore, the zero dynamics (2.20) are unstable and the flexible–link system is non–minimum phase. Consequently, the system cannot be inverted and the inverse dynamics control strategy results in an unstable closed–loop system. Therefore, for such a system, perfect or asymptotic convergent tracking should not be pursued. Instead, one should find controllers which lead to acceptably small tracking errors for the desired trajectories of interest.

2.4.   Model Verification

In this section, the validity of the model represented by (2.6) is examined by comparing the responses of the system obtained by solving (2.6) to those obtained by the experimental manipulator. The experimental system consists of a single flexible–link whose parameters are taken from [91] and are shown in Table 2.1. In this table, l is the length of the link, γ is the mass per unit length, I_h is the hub inertia, b is the viscous friction at the hub, C_{coul} is the coefficient of Coulomb friction, E is Young's modulus, I is the beam area moment of inertia, ω_j is the jth resonance frequency of the beam, c_i's are the viscous damping coefficients, and M_l is the payload. Two flexible modes are considered in this model. More details about the experimental manipulator are given in Appendix C.

The validity of the model is tested by applying different patterns of torque signals to the experimental test–bed. The model responses to the same torques are then obtained and the accuracy of the model is investigated by comparing these two sets of responses.

Table 2.1. Link parameters for the experimental manipulator

l	1.2 m
γ	1.2 kg/m
I_h	.3 $kg.m^2$
b	.59 $N.m/rad.s^{-1}$
C_{coul}	$\dot{\theta} > 0$ 4.74 $N.m$
	$\dot{\theta} < 0$ 4.77 $N.m$
EI	1.94 $N.m^2$
ω_1	3 rad/s
ω_2	19 rad/s
c_1	0.4
c_2	4.0
M_l	30 g

The torque $\tau_1(t)$ and its responses are shown in Figures 2.2–a to 2.2–d. Figure 2.2–a shows the applied torque $\tau_1(t)$. The hub position, tip deflection and net tip position are shown in Figures 2.2–b to 2.2–d, respectively. Each figure includes the experimental result (solid line) and three simulation results which are obtained using different sets of Coulomb friction coefficients shown in Table 2.2. Specifically, in Figure 2.2, dashed lines correspond to C_{coul1}, dash-dot lines correspond to C_{coul2}, and dotted lines correspond to C_{coul3}. As can be seen, the best agreement can be obtained by setting the value of the Coulomb friction coefficients to C_{coul1}. This set of Coulomb friction coefficients differs from the values of Table 2.1. These changes are required because the original values are averaged over a range of hub angles and therefore yield imprecise responses when compared with the experimental responses, particularly for the hub position. In Figure 2.3, the torque $\tau_2(t)$ and its responses are shown. In these figures, each figure includes the simulation results (solid line) obtained by setting $C_{coul} = C_{coul1}$ and three experimental responses. As can be observed, the hub position response (Figure 2.3–b) is different for each experiment. These differences are caused by slight variations of the initial hub positions and demonstrate the variation of the actual Coulomb friction with hub position.

To alleviate this problem, the torque $\tau_3(t)$ is applied to the system.

Table 2.2. Different sets of Coulomb friction coefficients.

	$\dot{\theta} > 0$	$\dot{\theta} < 0$
C_{coul1}	4.8	4.55
C_{coul2}	4.95	4.6
C_{coul1}	4.95	4.55

Figure 2.4–a to 2.4–d show $\tau_3(t)$ and its responses. Each figure includes the experimental response (solid line) and three simulation responses obtained by setting $C_{coul} = C_{coul1}$ (dashed line), $C_{coul} = C_{coul2}$ (dotted line) and $C_{coul} = C_{coul3}$ (dash-dot line). Figure 2.4–b shows the hub position responses, and a considerable amount of error can be observed for the case $C_{coul} = C_{coul1}$ (dashed line) which yields close agreement when applying torque $\tau_1(t)$. In this case, the best result is obtained by setting $C_{coul} = C_{coul2}$ (dotted line). Consequently, the use of a single set of Coulomb friction coefficients leads to an inaccurate model. Neglecting the stiction also contributes to inaccuracy. The inaccuracy is particularly significant when the value of the applied torque approaches that of the Coulomb friction.

To verify the linear model of the flexible–link manipulator, the Coulomb friction at the hub should be canceled. This cannot be done in an open–loop manner since the exact value of Coulomb friction coefficient is not known and it varies with hub position. To resolve this problem, the Coulomb friction can be compensated for in a closed–loop system. A PD control strategy with very high gains [91] is able to overcome the effect of the friction. The linear model of the system can then be verified by comparing the closed–loop responses of the model to those of the experimental system.

Towards this end, a reference trajectory $\theta_r(t)$ (Figure 2.5–a) is selected for this experiment. Appropriate selection of controller gains K_p and K_v ensures that torque saturation does not occur. A proportional gain K_p of 3000 and a derivative gain K_v of 25 ensure slightly over-damped tracking of $\theta_r(t)$. Figure 2.5–a to 2.5–d show the experimental responses (dashed lines) and simulation responses (solid lines) to the reference trajectory $\theta_r(t)$. Figure 2.5–a shows the hub position responses, and close agreement between the simulation and experimental responses can be observed. The tip deflection is shown in Figure 2.5–b. Note that until $t = 7$ seconds, the peak amplitudes of the experimental curve are greater than those of the simulated plot. This behavior is a result of

Table 2.3. Poles and zeros corresponding to the tip position

Poles	Zeros
0	13.8212
-0.6355	30.5486
$-1.1531 \pm 6.1678j$	-9.6169
$-1.5910 \pm 22.6452j$	-33.4977

the linearly decaying response that is a characteristic of the Coulomb friction present in the actual test–bed and of the exponentially decaying response characteristic of the viscous damping used in the model. The tip position responses are shown in Figure 2.5–c, and the input torque $\tau(t)$ is shown in Figure 2.5–d and is within saturation limits of ± 35.25 $N.m$

The above analysis demonstrates that the linear model (2.14) approximates the actual system reasonably well in the absence of Coulomb friction and is suitable for defining the new output and designing the controllers. The former is the subject of the next chapter.

2.4.1. A Linear Model of the Experimental Manipulator

The linearized state–space equations (2.15) can be transformed to the transfer function representation

$$G_t(s) = \frac{y_t(s)}{u(s)} = \frac{p_t(s)}{q(s)}, \qquad (2.22)$$

where $p_t(s)$ and $q(s)$ are polynomial functions of the Laplace transform variable s, and $y_t(s)$ and $u(s)$ are the Laplace transforms of $y_t(t)$ and $u(t)$, respectively. For the experimental manipulator whose parameters are given in Table 2.1,

$$p_t(s) = 0.16s^4 - 0.2s^3 - 1.88 \times 10^2 s^2 + 6.28 \times 10^2 s + 2.18 \times 10^4,$$
$$q(s) = s^6 + 6.12s^5 + 5.65 \times 10^2 s^4 + 16.71 \times 10^3 s^3 + 2.11 \times 10^4 s^2$$
$$+ 1.29 \times 10^4 s.$$

The poles and zeros of $G_t(s)$ are simply the roots of $q(s)$ and $p_t(s)$, respectively which are given in Table 2.3 and are plotted in Figure 2.6. The presence of two RHP zeros (see Table 2.3) identifies $G_t(s)$ as a non–minimum phase transfer function. Consequently, the inverse dynamics

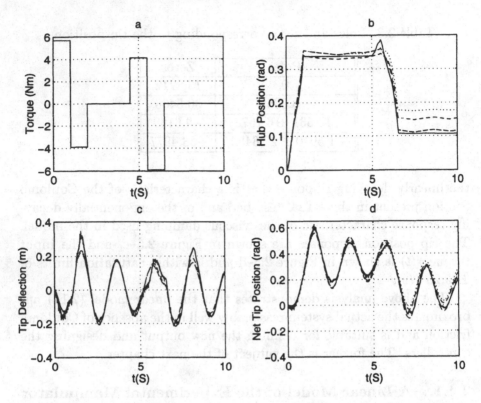

Figure 2.2. Open–loop responses to the applied torque $\tau_1(t)$: (a)- Applied torque $\tau_1(t)$, (b) Experimental (solid) and simulated hub angles $\theta(t)$, (c) Experimental (solid) and simulated tip deflections $W(l,t)$, (d) Experimental (solid) and simulated net tip positions $y_t(t)$.

control strategy based on the tip–position as the system output cannot be applied to control the flexible–link system. In the following chapter, the output of the system is redefined such that the corresponding zero dynamics are stable. This output is then used for performing the inverse dynamics control discussed in later chapters.

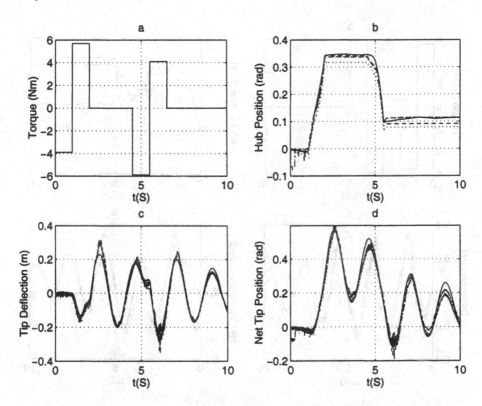

Figure 2.3. Open–loop responses to the applied torque $\tau_2(t)$: (a)- Applied torque $\tau_2(t)$, (b) Simulated (solid) and experimental hub angles $\theta(t)$, (c) Simulated (solid) and experimental tip deflections $W(l,t)$, (d) Simulated (solid) and experimental net tip positions $y_t(t)$.

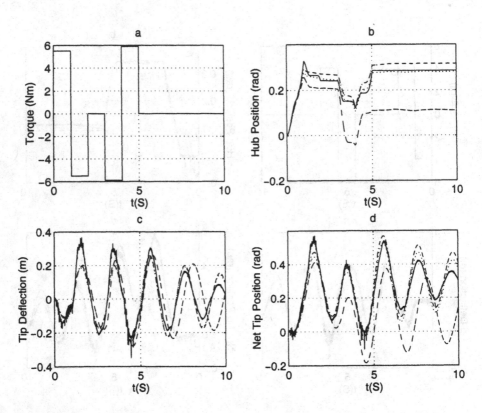

Figure 2.4. Open–loop responses to the applied torque $\tau_3(t)$: (a)- Applied torque $\tau_3(t)$, (b) Experimental (solid) and simulated hub angles $\theta(t)$, (c) Experimental (solid) and simulated tip deflections $W(l, t)$, (d) Experimental (solid) and simulated net tip positions $y_t(t)$.

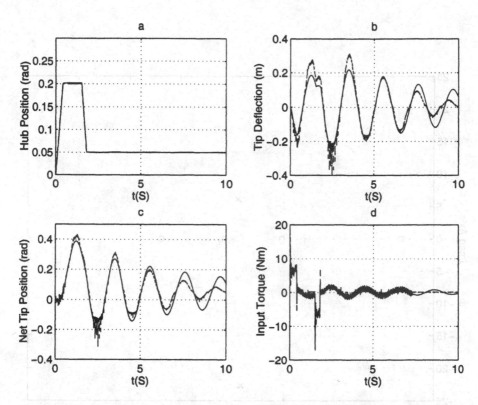

Figure 2.5. Closed–loop responses for the PD hub control: (a)- Desired hub position $\theta_0(t)$, simulated hub position (solid) and experimental hub position $\theta(t)$ (dashed), (b) Experimental (dashed) and simulated tip deflection $W(l,t)$ (solid), (c) Experimental (dashed) and simulated net tip positions $y_t(t)$ (solid), (d) Experimental (dashed) and simulated control torque $u(t)$ (solid).

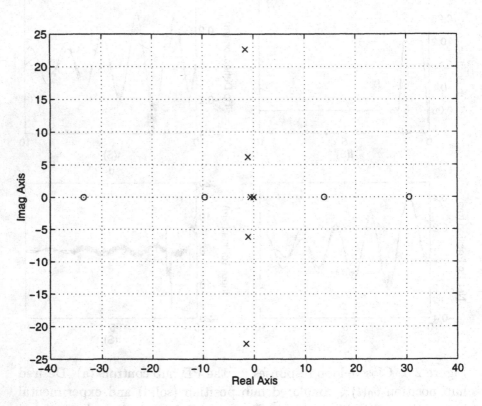

Figure 2.6. Pole–zero map of the tip position transfer function $G_t(s)$.

Chapter 3

Output Redefinition

In this chapter, the output redefinition method used for the first two control schemes proposed in this monograph is explained. First, a brief introduction is given regarding the output redefinition approach for non-linear systems. Then, application of this idea to control a flexible–link manipulator is discussed. Finally, this approach is modified to enable the designer to employ this scheme with minimum *a priori* knowledge about the system dynamics. Towards this end, a linear model of the system is assumed to be known with no *a priori* knowledge about the payload mass.

3.1.   Introduction

As stated earlier, application of inversion control techniques for exact trajectory tracking of non–minimum phase systems is impossible with a bounded control input. Several methods have been proposed in the literature to control non–minimum phase nonlinear systems. One methodology is to bypass the minimum phase requirement by applying an input–state rather than input–output (inverse dynamics) linearization. Tornambe [92] studied output feedback stabilization for *observable* input–state linearizable nonlinear systems. The strategy is based on (*i*) introducing a cascade precompensator dynamics, (*ii*) input–state linearizing the augmented system, and (*iii*) converting linear state–feedback control into dynamical output feedback control using the observability property assumed for the system.

However, the input–state linearization approach is not suitable for output tracking control unless there is a way to express the desired states in terms of the desired output trajectory which is not always straightforward. Moreover, the flexible–link system is not input–state linearizable as shown by Wang and Vidyasagar [5].

In view of the potential advantages of an inversion control law, e.g. its straightforward extension to the nonlinear setting and motivated by the fact that the zero dynamics depend on the choice of the output, it may be convenient to slightly modify the problem specifications in order to achieve a minimum phase characteristic for the system. One interesting approach is the *output redefinition* method whose principle is to redefine the output function so that the resulting zero dynamics are stable. Then an inverse dynamics control strategy can be designed based on the new output. However, the important question to be answered is the following: Does the actual output track the desired trajectory as closely as possible? There are two ways to deal with this question.

Gopalswamy and Hedrick [58, 93] proposed a sliding control strategy based on (*i*) defining a new output such that the associated zero dynamics are asymptotically stable and (*ii*) defining a modified desired trajectory such that asymptotic tracking of the modified desired trajectory by the new output results in asymptotic tracking of the original desired trajectory by the original output. However, modifying the desired trajectory is not possible for every problem.

Consequently, approximate tracking control has been proposed in the literature. In this way, the output of the nonlinear system is redefined in such a way that it is essentially the same as the original output in the frequency range of interest, Hence, exact tracking of the new output also implies approximate tracking of the original output. When performing input–output linearization using successive differentiations of the output, one practical approximation [61] is to simply *neglect* the terms containing the input and keep differentiating the selected output a number of times equal to the system order, so that there are "approximately" no zero dynamics. Of course, this approach can only be meaningful if the coefficients of u at the intermediate steps are "small" , *i.e*, if the systems are "weakly" non–minimum phase. The approach is conceptually similar to neglecting "fast" right–half plane zeros in linear systems (in the frequency domain, $1 - \tau s \cong 1/(1 + \tau s)$ if $|\tau s| \ll 1$ *i.e*, if the zero $(1/\tau)$ is much faster than the frequency range of interest).

3.2. Redefining the Output

It is well–known that the zero dynamics of a flexible–link manipulator associated with the tip position are unstable. In other words, the system is non–minimum phase and direct application of the inverse dynamics control strategy results in unstable closed–loop internal dynamics. Wang and Vidyasagar [44] proposed the reflected tip position, i.e. $y_{ri} = \theta_i - \frac{W_i(l_i,t)}{l_i}$, where it can be shown that the zero dynamics related to this output are stable, and consequently the system is minimum phase and can be stabilized by a PD-type control law. This output is easy to compute from the hub and tip position measurements i.e. $y_{ri} = 2\theta_i - y_{ti}$. The main advantage of using the reflected-tip position control over the joint-based control is that by using a joint-based control strategy the vibrations of the system cannot be controlled, and as a result the only damping experienced by the system is its natural damping. Therefore, the vibrations of the elastic modes take a long time to die out resulting in considerable oscillations at the tip. Note that, as the speed of the reference trajectory is increased, the unmodeled higher frequency flexible modes will become excited. Given that $y_{ti} = \theta_i + \frac{W_i(l_i,t)}{l_i}$ and $y_{ri} = \theta_i - \frac{W_i(l_i,t)}{l_i}$, therefore the difference between the reflected–tip position (RTP) and the actual tip position becomes significant for high–speed reference trajectories and hence acceptable tracking performance cannot be ensured. The same argument applies to the case when the link under control is very flexible and the flexible modes are excited easily. This issue will be described further in Chapter 4. Hence, for a very flexible system instead of RTP, a different output should be defined for designing the control law.

De Luca and Lanari [47] suggested a set of actuation and sensing points which convert the input-output mapping to a minimum phase one. The more natural choice of keeping the actuation point at the joint while varying the output location along the link has been considered in [48]. An alternative approach is to keep the output fixed at the tip and let the actuation point vary along the link [94]. These methods however, suffer from the point of view of practical applications, i.e., the installation of the actuator or sensor along the link.

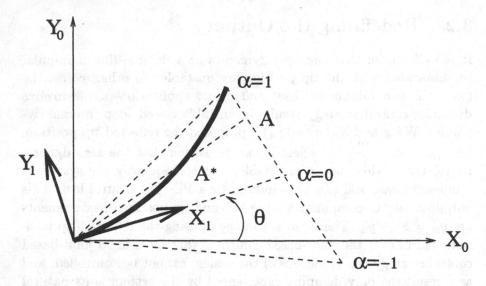

Figure 3.1. Outputs of the flexible link.

3.2.1. Defining an Output Close to the Tip Position

To define the new output as close as possible to the end-effector, Madhavan and Singh [49] used the joint angle plus a scaling of the tip elastic deformation as the output for control in each link, namely $y_{ai} = \theta_i + \alpha_i \frac{W_i(l_i,t)}{l_i}$, where $-1 < \alpha_i < 1$ (see Figure 3.1). It can be seen that different values of α correspond to different points on the beam. As an example, the point A^* on the link corresponds to a particular value of α. Therefore, control of the point A accomplishes the control of point A^*. For the choice of $\alpha = 1$, the output becomes the tip angular position, for $\alpha = 0$ the output becomes the joint angle, and for $\alpha = -1$ the output becomes RTP.

In [49], the authors also showed that a positive critical value $\alpha_i^* < 1$ exists such that the zero dynamics related to the new output, y_{ai} are unstable for all $\alpha_i > \alpha_i^*$ and are stable for all α_i satisfying $-1 < \alpha_i < \alpha_i^*$. Hence an inverse dynamics controller can be designed to control the system output for α_i in the range $-1 < \alpha_i < \alpha_i^*$. Even though one must choose $\alpha_i = 1$ for the tip position control, in order to avoid unstable zero dynamics, one must keep $\alpha_i < \alpha_i^* < 1$. This evidently leads to trajectory control of a coordinate close to the actual tip position. Our objective in this section is to show that by using the new output y_{ai}, the dynamics of the flexible-link manipulator may be expressed such that the feedback–error learning method (see Chapter 4) is applicable for controlling the system.

Consider the dynamics of the manipulator given by equation (2.12) and define $H(\theta, \delta) = M^{-1}(\theta, \delta) = \begin{bmatrix} H_{11} & H_{12} \\ H_{21} & H_{22} \end{bmatrix}$. Then (2.12) can be re–written as

$$\begin{bmatrix} \ddot{\theta} \\ \ddot{\delta} \end{bmatrix} = H(\theta, \delta) \begin{bmatrix} u - f_1(\theta, \dot{\theta}) - h_1(\theta, \dot{\theta}, \delta, \dot{\delta}) - F_1 \dot{\theta} - f_c \\ -f_2(\theta, \dot{\theta}) - h_2(\theta, \dot{\theta}, \delta, \dot{\delta}) - K\delta - F_2 \dot{\delta} \end{bmatrix} \qquad (3.1)$$

Defining $\Gamma_{n \times m} = \Upsilon \gamma$, where $\Upsilon = diag\{\alpha_1 \cdots \alpha_n\}$ and γ is given by (2.13), the new output can be expressed as $y_a = \theta + \Gamma_{n \times m} \delta$. Now, consider system (2.12) with the output defined above. To find the external dynamics related to this new output successive time differentiation have to be taken until the input appears, namely

$$\ddot{y}_a = \ddot{\theta} + \Gamma_{n \times m} \ddot{\delta} \qquad (3.2)$$

Using (3.1) and (3.2), it follows that

$$\ddot{y}_a = A(\theta, \dot{\theta}, \delta, \dot{\delta}) + B(\theta, \delta)u \qquad (3.3)$$

where

$$B(\theta, \delta) = H_{11} + \Gamma_{n \times m} H_{21}$$

and

$$\begin{aligned} A(\theta, \dot{\theta}, \delta, \dot{\delta}) &= -(H_{11} + \Gamma H_{21})(f_1 + h_1 + F_1 \dot{\theta} + f_c) \\ &\quad - (H_{12} + \Gamma H_{22})(f_2 + h_2 + K\delta + F_2 \dot{\delta}) \end{aligned}$$

The external dynamics related to the new output can be written in general as

$$u = f(\theta, \dot{\theta}, \delta, \dot{\delta}, \ddot{y}_a) \qquad (3.4)$$

Zero dynamics by definition in [95] are the dynamics which are left in the system once the input is chosen in such a way that it constrains the output to remain at zero. This input can be obtained from (3.3) as

$$u = B^{-1}(\theta, \delta)[-A(\theta, \dot{\theta}, \delta, \dot{\delta})]$$

The zero dynamics of the system may now be expressed as

$$\ddot{\delta} = -P[f_2(w1, w2) + h_2(w1, w2, w3, w4) + K\delta + F_2 \dot{\delta}] \qquad (3.5)$$

where

$$w1 = -\Gamma\delta, \quad w2 = -\Gamma\dot{\delta}, \quad w3 = \delta, \quad w4 = \dot{\delta}$$

and P is given by

$$P = [H_{22} - H_{21}(H_{11} + \Gamma H_{21})^{-1}(H_{12} + \Gamma H_{22})]\,|_{(w1,w3)} \qquad (3.6)$$

At this stage by linearizing the zero dynamics, one can find the value of α_i^*. But, since the mass matrix M (and hence H and P) depends on the payload M_l, α_i^* also depends on the payload mass. Consequently, to obtain the exact value of α_i^*, the value of M_l should be known *a priori*. But, the control schemes proposed in this monograph assume no *a priori* knowledge about the payload mass M_l. In Section 3.3, it will be shown that the dependence of α_i^* to the payload mass M_l is such that the value of α_i^* takes its lowest value when M_l is zero. In other words, the value of α_i^* obtained for zero payload mass guarantees stability of the zero dynamics as M_l increases. This choice of α_i^* is conservative, since as M_l increases larger values of α_i^* can be used.

Now, by neglecting the payload (for the purpose of performing output redefinition only), and by linearizing (3.5), equations (3.5) and (3.6) become

$$\ddot{\delta} = -P_0[K\delta + F_2\dot{\delta}] \qquad (3.7)$$
$$P_0 = [H_{22} - H_{21}(H_{11} + \Gamma H_{21})^{-1}(H_{12} + \Gamma H_{22})]\,|_{(0,0)}$$

where P_0 is the value of the linearized P evaluated at $M_l = 0$. Now, suppose that the vector α and the matrices H, K, and F_2 are such that

$$A(\Upsilon) = \begin{bmatrix} 0 & I \\ -P_0 K & -P_0 F \end{bmatrix}$$

is Hurwitz. Then the origin of (3.7), and hence (3.5), is locally asymptotically stable and the original nonlinear system is locally minimum phase [6]. Provided that the linearized mass matrix M (with zero payload), the stiffness matrix K, and the viscous damping matrix F_2 are known, then a proper output may be specified by obtaining an Υ such that $A(\Upsilon)$ is guaranteed to be Hurwitz. The variation of α_i^* to changes in the viscous damping matrix F_2 are discussed in Appendix A.

3.3.   Variation of α_i^* with Payload

In the previous section, an output redefinition approach for flexible–link manipulators has been presented based on zero payload mass. In the following, it is shown that using the value of α_i^* obtained for $M_l = 0$ ensures stability of the zero dynamics as payload mass increases.

Towards this end, consider the dynamics of a single flexible–link manipulator when one flexible mode is considered. The dynamic equations of the manipulator is given by (2.16) and is repeated here as

$$
\begin{aligned}
m_{11}\ddot{\theta} + m_{12}\ddot{\delta} + F_1\dot{\theta} &= u \\
m_{12}\ddot{\theta} + m_{22}\ddot{\delta} + F_2\dot{\delta} + K\delta &= 0.
\end{aligned}
\qquad (3.8)
$$

It is assumed that there is an $\alpha^* = \alpha_0$ which ensures stability of the zero dynamics associated with the new output for $M_l = 0$. When nonzero payload is considered in the model, $m_{ij}, \; i, j \in \{1, 2\}$ can be obtained from (2.8) as

$$
\begin{aligned}
m_{11} &= J_0 + I_h + M_l l^2, \\
m_{12} &= \rho A \int_0^l \phi_1(x)x\,dx + M_l l\phi_{1e}, \\
m_{22} &= \rho A + M_l \phi_{1e}^2,
\end{aligned}
\qquad (3.9)
$$

and K and F_2 are given by (2.9) and (2.10), respectively. Now, defining the new output as

$$
y_a = \theta + \alpha_0 \frac{\phi_{1e}}{l}\delta,
$$

where α_0 is the value of α^* obtained for $M_l = 0$, and following the same procedure as in Section 2.3, the zero dynamics of the system can be expressed as

$$
(m_{22} - \alpha_0\frac{\phi_{1e}}{l}m_{12})\ddot{\delta} + F_2\dot{\delta} + K\delta = 0
\qquad (3.10)
$$

The stability of the zero dynamics (3.10) can be investigated by specifying the sign of the coefficient $(C_{z\alpha} = m_{22} - \alpha_0\frac{\phi_{1e}}{l}m_{12})$. Using (3.9), $C_{z\alpha}$ can be expressed as

$$
\begin{aligned}
C_{z\alpha} &= \rho A + M_l\phi_{1e}^2 - \alpha_0\frac{\phi_{1e}}{l}(\rho A \int_0^l x\phi_1(x)dx + M_l l\phi_{1e}) \\
&= \rho A(1 - \alpha_0\frac{\phi_{1e}}{l}\int_0^l x\phi_1(x)dx) + (M_l\phi_{1e}^2 - \alpha_0\frac{\phi_{1e}}{l}M_l l\phi_{1e}).
\end{aligned}
\qquad (3.11)
$$

In (3.11), $C_{z\alpha}$ can be written as $C_{z\alpha} = C_{z\alpha 0} + C_{z\alpha m}$ where

$$C_{z\alpha 0} = \rho A(1 - \alpha_0 \frac{\phi_{1e}}{l} \int_0^l x\phi_1(x)dx)$$

represents $C_{z\alpha}$ when $M_l = 0$, and

$$C_{z\alpha m} = M_l \phi_{1e}^2 - \alpha_0 \frac{\phi_{1e}}{l} M_l l \phi_{1e} \tag{3.12}$$

represents the terms depending on the payload.

It is assumed that α_0 is the value of α^* obtained for $M_l = 0$. This implies stability of the zero dynamics for the zero payload case, and hence $C_{z\alpha 0} > 0$. Now, Consider $C_{z\alpha m}$ in (3.12) which can be expressed as

$$C_{z\alpha m} = M_l \phi_{1e}^2 - \alpha_0 M_l \phi_{1e}^2 = M_l \phi_{1e}^2 (1 - \alpha_0). \tag{3.13}$$

It can be concluded from (3.13) that $C_{z\alpha m} > 0$ since $|\alpha_0| < 1$ and M_l is always positive. Consequently, $C_{z\alpha} > 0$ and stability of the zero dynamics for this output is always ensured as M_l increases. Hence, an output redefinition can be performed without *a priori* knowledge about the payload mass M_l.

3.4.   Conclusions

In this chapter, a modified output redefinition approach was proposed that requires only *a priori* knowledge about the linear model of the system and no *a priori* knowledge about the payload mass. First, the output of the flexible–link system was redefined such that the zero dynamics related to the new output are stable. This is done by neglecting the payload and linearizing the zero dynamics of the system. Then, the stability of the zero dynamics associated to this new output was shown to be ensured as payload mass M_l increases. This enables us to design such controllers that remain robust to the payload variation. This is the subject of the next chapter.

Chapter 4

Proposed Neural Network Structures

In this chapter, different control and neural network strategies are proposed for tip position tracking control of the flexible–link manipulators using the new output approach discussed in the previous chapter. In Section 4.1, an introduction is provided regarding learning control using neural networks. Then, control strategies are presented in the subsequent sections. Simulation results for two single flexible–link manipulators and a two–link manipulator are presented in Section 4.6. Section 4.7 discusses design and training issues for neural networks.

4.1. Introduction

4.1.1. Function Approximation for Neural Networks

A number of diverse application areas are concerned with the representation of general functions of an n–dimensional real variable, $\mathbf{x} \in \Re^n$, by finite linear combinations of the form

$$\sum_{j=1}^{N} v_j \varsigma(\mathbf{w}_j^T \mathbf{x} + \chi_j),$$

where $\mathbf{w}_j \in \Re^n$ and $v_j, \chi_j \in \Re$ are fixed. Here the univariate function ς depends heavily on the context of the application. The major concern with the so–called sigmoidal ς is:

$$\varsigma(t) \to \begin{cases} 1 & \text{as } t \to +\infty, \\ 0 & \text{as } t \to -\infty \end{cases} \tag{4.1}$$

Such functions arise naturally in neural network theories as the activation function of a neural node.

Neural network capability to approximate nonlinear functions has been investigated by several researchers [96, 97, 98, 99]. They have shown that a wide range of nonlinear functions can be approximated arbitrarily closely by a feedforward neural network. Cybenko [96] has shown the following:

Theorem 4.1 *Let I_n denote the n-dimensional unit cube, $[0,1]^n$. The space of continuous functions on I_n is denoted by $C(I_n)$. Let ς be any continuous sigmoidal function of the form (4.1). Then finite sums of the form*

$$G(\mathbf{x}) = \sum_{j=1}^{N} v_j \varsigma(\mathbf{w}_j^T \mathbf{x} + \chi_j),$$

are dense in $C(I_n)$. In other words, given any $f \in C(I_n)$ and $\epsilon > 0$, there is a sum, $G(\mathbf{x})$, of the above form, for which

$$\mid G(\mathbf{x}) - f(\mathbf{x}) \mid < \epsilon \quad \text{for all} \quad \mathbf{x} \in I_n.$$

Various neural network methods have been proposed in recent years, but the one most relevant to control is the well-known backpropagation method. Although many interpretations have been given to backpropagation networks such as "perception", "recognition", "internal representation", "encoding", and so on, neural network control schemes are mostly based on interpreting the backpropagation network as a method of function approximation. This interpretation not only demonstrates the sound mathematical foundation of the backpropagation network, but also allows a simple intuitive understanding of its capabilities. Hecht-Nielsen [98] has obtained the following result:

Theorem 4.2 *Given any $\epsilon > 0$ and any L_2 function $\mathbf{f} : [0,1]^n \in \Re^n \to \Re^m$, there exists a three-layer backpropagation network that can approximate $\mathbf{f}$ to within ϵ mean-squared error accuracy.*

Although a three-layer backpropagation network has been shown to be capable of approximating any arbitrary function, the important questions that remain to be answered deals with how many neural nodes are required to yield an approximation of a given function? What properties of the function being approximated play a role in determining the number of neurons? As a first try, one can use Kolmogorov's theorem [100] to get the following result:

Theorem 4.3 *Given any continuous function* $f : [0,1]^n \in \Re^n \to \Re^m$, *there exists a three-layer feedforward network having n fan-out neurons in the first layer, $(2n + 1)$ neurons in the hidden layer, and m neurons in the output layer which can approximate f to any desired degree of accuracy.*

However, Theorem 4.3 does not say how to obtain the activation functions for which the function can be approximated by $(2n+1)$ neurons in the hidden layer. Consequently, this theorem cannot be considered as a generic rule and the size of the network has to be decided depending on the problem.

4.1.2. Learning Control using Neural Networks

A number of techniques can be used to design controllers for unknown linear systems. Typically, a standard model structure is used and then the parameters of controllers or plant models are adapted based on stability theory. On the other hand, the control of uncertain nonlinear systems is difficult for a number of reasons. First, it is not easy to find a suitable model structure for the nonlinear dynamics unlike linear systems where a standard form of the transfer function is available for an unknown system of a given order. Secondly, there is no standard way of generating adaptation laws for nonlinear systems.

The nonlinear mapping properties of neural networks are central to their use in control engineering. Feedforward networks, such as the Multilayer Perceptron can be readily thought of as performing an adaptive nonlinear vector mapping. Adaptive neural controllers can be roughly categorized according to the means by which the controller parameters are adjusted. The two common strategies are direct and indirect adaptive controllers. In indirect adaptive controllers [101, 102, 103], a network is first trained to identify input–output behavior of the plant. Using the resulting identification model, the controller is designed based on a cancellation scheme.

The general structure for a direct adaptive controller is shown in Figure 4.1 [104]. For this adaptive control problem formulation, it is necessary to adjust the weights of the neural network during the learning phase to produce a nonlinear controller that can control the nonlinear plant in such a manner that a cost function of the plant output and the desired response is minimized. Due to the nonlinearities present in both the plant and the controller, stability based adaptation laws which are

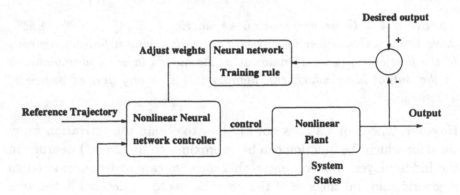

Figure 4.1. Schematic of the direct neural adaptive controller

widely used in linear adaptive control, have not as yet been proposed for the neural adaptive controller. The early research on neural adaptive control is based on gradient descent techniques in which the cost function can be minimized by the adaptation of the network weights in the negative direction of the gradient of the cost with respect to these weights. A common algorithm is the backpropagation rule that provides the necessary gradient of the cost function with respect to each weight.

In Figure 4.1, the plant is situated between the neural network and the error. Hence, this error cannot be directly used to adjust the controller parameters and it is necessary to find some method by which the error at the output of the plant can be fed back to produce a suitable descent direction at the output of the neural network. Psaltis and Sideris [105], introduced the concept of using the plant Jacobian to allow errors at the plant output to be fed back to the network. If the cost function is defined as $J(\mathbf{w})$ where $\mathbf{w}$ is the matrix of the weights of the neural network, then by knowing the Jacobian of the plant, the gradient of the cost function with respect to the jth input u_j can be readily determined with y_i being the ith plant output that is

$$\frac{\partial J(\mathbf{w})}{\partial u_j} = \sum_{i=1}^{n} \frac{\partial J(\mathbf{w})}{\partial y_i} \frac{\partial y_i}{\partial u_j}$$

The plant is considered as an additional unmodifiable layer. Therefore, the error signal at the system output can be propagated back through the plant using the partial derivatives of the plant at its operating point. Then the backpropagation rule can be used to adjust the weights in all layers except the virtual output layer (the plant). However, since it is to be assumed that little knowledge of the nonlinear plant is available, it is difficult to obtain an analytical expression for the plant Jacobian. There are a number of methods by which the problem of backpropagating the error through the plant to the controller can be solved.

Saerens and Soquet [106], suggested the use of the sign of the Jacobian instead of its real value for the training of neural adaptive controllers. This is often available simply from qualitative knowledge of the system. The plant backpropagation equation then becomes:

$$\frac{\partial J(\mathbf{w})}{\partial u_j} \cong \sum_{i=1}^{n} \frac{\partial J(\mathbf{w})}{\partial y_i} \, SGN\left(\frac{\partial y_i}{\partial u_j}\right) \tag{4.2}$$

It should be noted that the scalar product between the gradient produced by the true method and that produced by the approximation is always positive and will hence ensure error minimization.

Nguyen and Widrow [107] and Jordan and Jacobs [108] proposed using a neural forward model of the plant as a channel for the backpropagation of errors to the neural controller. A neural network is first trained to provide a model of the nonlinear plant. This can then be used in parallel with the plant with errors at the plant output backpropagated through the model to form the necessary gradients at the output of the neural controller, hence avoiding the need to know the plant Jacobian.

4.2. Feedback–error Learning

Feedback–error learning was first introduced in [109] and later was modified in [110]. Two adaptive learning control schemes were proposed using feedback–error learning for neural network controllers. In both schemes, the error signal for the neural network is the output of the feedback controller. The advantage of this learning scheme is that the target signal or the desired output signal for the neural network is not required. Also, back propagation of the error (training) signal through the controlled system or through the model of the controlled system is not required. Since the feedback torque is chosen as the error signal for this learning strategy, the feedback torque is expected to decrease with learning. This implies that the error between the desired trajectory and the actual trajectory is required to tend to zero as learning evolves. The learning scheme is designated as "feedback–error learning"

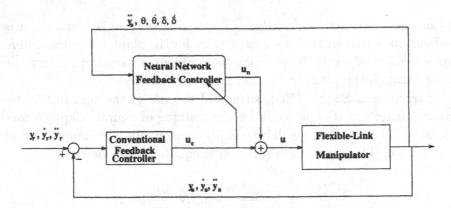

Figure 4.2. Structure of Inverse Dynamic Model Learning (IDML).

to emphasize the importance of using the feedback torque as the error signal.

The idea of feedback–error learning has been applied to control a rigid-link robot manipulator in [110, 109] where the system has no zero dynamics and all the state variables are available. For a flexible-link manipulator, however, the zero dynamics related to the tip are unstable and full state feedback is not available. As mentioned earlier, a certain output can be defined so that the zero dynamics corresponding to this new output are stable. This output can be defined as the joint variable, but it will not yield an acceptable tip response for a relatively flexible robot. In this section, it is shown that by using the new output that was defined in the previous chapter, the concept of feedback–error learning may be invoked to control the tip position of a flexible-link manipulator.

4.2.1.   Inverse Dynamics Model Learning (IDML)

The structure of the first scheme is referred to as Inverse Dynamics Model Learning (IDML) and is shown in Figure 4.2. The system dynamics are assumed to be governed by

$$f(\theta, \dot{\theta}, \delta, \dot{\delta}, \ddot{y}_a) = u$$

In this scheme, a conventional feedback controller (CFC) and a neural network controller are connected as depicted in Figure 4.2. The CFC is used both as an ordinary feedback controller to guarantee asymptotic stability of the system during the learning period and as a "reference model" for the response of the controlled system. For example, a linear controller can be expressed by the following equation

$$u_c = K_2(\ddot{y}_r - \ddot{y}_a) + K_1(\dot{y}_r - \dot{y}_a) + K_0(y_r - y_a) \qquad (4.3)$$

where $y_r, \dot{y}_r, \ddot{y}_r$ denote the desired trajectory (i.e., position, velocity and acceleration, respectively). The goal of the neural network feedback controller is to ultimately represent the inverse dynamics model of the controlled system. The output of the CFC is fed to the network for the error signal. The neural network also receives $\theta, \dot{\theta}, \delta, \dot{\delta}, \ddot{y}_a$ as ordinary inputs. The output of the network is

$$u_n = \Phi(\theta, \dot{\theta}, \delta, \dot{\delta}, \ddot{y}_a, \mathbf{w})$$

where $\mathbf{w}$ is the matrix of the weights of the neural network. The neural network can be one of several types of neural network models in which the error of the output vector will decrease by changing the internal adaptive parameters (the weights). The assumption about the network is that the nonlinear function of the controlled object, can be arbitrary modeled closely by Φ with an appropriate $\mathbf{w}$ within a compact set. The learning rule specified for the feedback–error learning scheme is given by

$$\dot{\mathbf{w}} = \eta \frac{\partial \Phi}{\partial \mathbf{w}} u_c \qquad (4.4)$$

where η is the learning rate. After learning, the neural network acquires an "arbitrarily" close model of the inverse dynamics of the controlled system, and the response of the controlled system is now governed by

$$K_2 \ddot{e} + K_1 \dot{e} + K_0 e \cong 0$$

where $e = y_r - y_a$. That is to say, the output tracking error, e, converges to zero in accordance with the above reference model.

4.2.2. Nonlinear Regulator Learning (NRL)

The configuration of the second learning scheme, Nonlinear Regulator Learning (NRL) is shown in Figure 4.3. Consider the dynamics of the flexible-link system (3.3). Multiplying (3.3) by $B^{-1}(\theta, \delta)$, we get

$$R(\theta, \delta)\ddot{y}_a + N(\theta, \dot{\theta}, \delta, \dot{\delta}) = u \qquad (4.5)$$

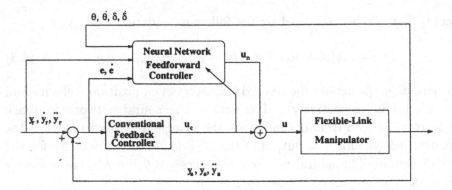

Figure 4.3. Structure of Nonlinear Regulator Learning (NRL).

where

$$R(\theta, \delta) = B^{-1}(\theta, \delta)$$
$$N(\theta, \dot{\theta}, \delta, \dot{\delta}) = -B^{-1}(\theta, \delta) A(\theta, \dot{\theta}, \delta, \dot{\delta}).$$

As with the IDML method, the CFC in Figure 4.3 serves the same two purposes. In comparison with the IDML case, the actual acceleration is not used as an input to the neural network in the NRL scheme. Instead, the reference trajectories (i.e. position, velocity, and acceleration) are fed to the neural network in order to generate the feedforward term so that better transient response is obtained at the early stage of the learning when the tracking error and hence u_c is large. The output of the neural network may be expressed as

$$u_n = \Omega(\ddot{y}_r, \dot{y}_r, y_r, y_r - y_a, \dot{y}_r - \dot{y}_a, \theta, \dot{\theta}, \delta, \dot{\delta}, \mathbf{w}) = \Phi(\ddot{y}_r, \dot{y}_r, y_r, \theta, \dot{\theta}, \delta, \dot{\delta}, \mathbf{w})$$

for some nonlinear functions Ω and Φ. The weights of the neural network are updated according to $\dot{\mathbf{w}} = \eta \frac{\partial \Phi}{\partial \mathbf{w}} u_c$. The closed-loop equation is obtained by applying $u = u_c + u_n$ to (4.5) to yield

$$(R(\theta, \delta) + K_2)(\ddot{y}_r - \ddot{y}_a) + K_1(\dot{y}_r - \dot{y}_a) + K_0(y_r - y_a) +$$
$$\Phi - N(\theta, \dot{\theta}, \delta, \dot{\delta}) - R(\theta, \delta)\ddot{y}_r = 0$$

If Φ can be made equivalent to Φ_d defined as

$$\Phi_d = N(\theta, \dot{\theta}, \delta, \dot{\delta}) + R(\theta, \delta)\ddot{y}_r + R(\theta, \delta)K_2^{-1}(K_1(\dot{y}_r - \dot{y}_a) + K_0(y_r - y_a)),$$

then the closed-loop dynamics may be expressed as

$$(R(\theta, \delta) + K_2)(\ddot{y}_r - \ddot{y}_a) + K_1(\dot{y}_r - \dot{y}_a) + K_0(y_r - y_a) +$$
$$R(\theta, \delta)K_2^{-1}(K_1(\dot{y}_r - \dot{y}_a) + K_0(y_r - y_a)) \cong 0$$

This gives

$$(I + R(\theta, \delta)K_2^{-1})(K_2(\ddot{y}_r - \ddot{y}_a) + K_1(\dot{y}_r - \dot{y}_a) + K_0(y_r - y_a) \cong 0$$

Consequently, provided that $I + R(\delta)K_2^{-1}$ is nonsingular in the above equation, then the tracking error dynamics become

$$K_2\ddot{e} + K_1\dot{e} + K_0 e \cong 0 \qquad (4.6)$$

Now by finding a proper α_i that ensures stability of the zero dynamics corresponding to the new output and by using full state feedback, the inverse dynamics strategy may be applied by utilizing the above two feedback–error learning schemes. However, it can be shown that for a single flexible–link manipulator these strategies require only measurements from the tip and the joint variables. Therefore the output $y_a = \theta + \alpha \frac{W(l,t)}{l}$ can be constructed from available measurements. Note that the tip position is given by $y_t = \theta + \frac{W(l,t)}{l}$; therefore $W(l,t)$ and $\dot{W}(l,t)$ may be obtained directly from y_t, θ, $\dot{y}_t$ and $\dot{\theta}$. Furthermore, it may also be shown that the nonlinear terms in the mass matrix $M(\delta)$ and in the Coriolis and the centrifugal forces $h_1(\dot{\theta}, \delta, \dot{\delta})$ and $h_2(\dot{\theta}, \delta)$ may be expressed as functions of $\dot{\theta}, W$, and $\dot{W}$. Towards this end, consider the terms in the dynamic equations of a single flexible–link manipulator given by (2.7) and (2.8)

$$M(\delta) = \begin{bmatrix} M_{11(1\times 1)}(\delta) & M_{12(1\times n)} \\ M_{21(n\times 1)} & M_{22(n\times n)} \end{bmatrix},$$

$$M_{11}(\delta) = m_0 + M_l(\phi^T \delta)^2$$

$$f_1 = 0$$

$$f_2 = 0$$

$$h_1(\dot{\theta}, \delta, \dot{\delta}) = 2M_l\dot{\theta}(\phi^T\delta)(\phi^T\dot{\delta})$$

$$h_2(\dot{\theta}, \delta) = -M_l\dot{\theta}^2(\phi\phi^T)\delta.$$

Now, using the definition of $W(l,t)$, M_{11} and $h(\dot{\theta}, \delta, \dot{\delta})$ may be expressed as

$$M_{11} = m_0 + M_l W(l,t)^2,$$

$$h(\theta, \dot{\theta}, \delta, \dot{\delta}) = \begin{bmatrix} 2M_l\dot{\theta}W(l,t)\dot{W}(l,t) \\ -M_l\dot{\theta}^2\phi W(l,t) \end{bmatrix}$$

Hence, except for the term $K\delta + F_2\dot{\delta}$ in (2.12) which is linear in δ and $\dot{\delta}$, all the nonlinearities may be expressed as functions of $\dot{\theta}, W$, and $\dot{W}$. In other words, approximate inverse dynamics may be achieved by providing $\dot{\theta}, W$, and $\dot{W}$ to the neural networks.

4.3.   Deflection Control for IDML and NRL

One of the main limitations of the output redefinition strategy is that the solution to α_i^* may become $\ll 1$. In other words, the point (output) under control could get too far away from the tip and too close to the hub. In this case, controlling the new output does not necessarily guarantee satisfactory response for the tip position due to the fact that there is no direct way to effectively damp out the elastic vibrations of the flexible modes at the tip. In [49], a linear stabilizer was employed that uses full state feedback and furthermore assumes linearity of the system dynamics close to the terminal phase of the desired trajectory. Our proposed approach attempts to overcome the above difficulty. Towards this end, we proposed to include the tip deflection directly in the objective function of the neural network. In other words, the error to the network is modified from u_c to $U_c = u_c + K_3 W(l, t)$, where u_c is given by (4.3) and $W(l, t)$ is the vector of deflection variables at the tip of each link. This amounts to modifying the objective function of the neural network to $J = \frac{1}{2}(e^T K_0 e + \dot{e}^T K_1 \dot{e} + \ddot{e}^T K_2 \ddot{e} + W(l, t)^T K_3 W(l, t))$ from $J = \frac{1}{2}(e^T K_0 e + \dot{e}^T K_1 \dot{e} + \ddot{e}^T K_2 \ddot{e})$. Consequently, direct control over the elastic vibrations of the flexible modes becomes feasible through K_3. The gain K_3 specifies the weight for controlling the vibrations where the tracking ability is specified by choosing u_c. Experimental results shown in Chapter 5 reveal that good control of the tip position can be obtained even when α_i^* is very small (i.e. close to and even equal to zero) and the link is "very" flexible.

4.4.   Joint–based Control (JBC)

In this section, the control structures developed in the previous section will be generalized by relaxing the a priori knowledge about the linear model of the flexible manipulator. Since a linear model of the system is not available, a PD-type control law cannot be designed to stabilize the system, and a suitable output for feedback cannot be determined as before. In other words, the feedback–error learning scheme cannot be

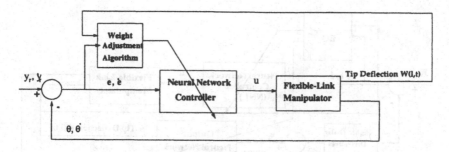

Figure 4.4. Joint–based neural controller.

applied directly. Instead, we adopt the general structure of an adaptive neural network which is shown in Figure 4.1.

The proposed control structure is shown in Figure 4.4. The key to designing this controller is to define the natural choice of joint position as the output for control which ensures the minimum phase property of the input–output map due to its colocated actuator/sensor pair and to damp out the tip elastic deformation by adding $W(l, t)$ in the cost function of the neural network. Consequently, the objective function for training the neural network is selected as $J = \frac{1}{2}(e^T K_1 e + \dot{e}^T K_2 \dot{e} + W(l, t)^T K_3 W(l, t))$, where $e := y_r - \theta$, and $W(l, t)$ is introduced to reduce the vibrations of the flexible modes of the system. The objective function J is a weighted function of e, $\dot{e}$ and $W(l, t)$ where the corresponding weights are specified by K_1, K_2 and K_3. Note that higher order derivatives of the error can also be included in the objective function. The inputs to the network are e, $\dot{e}$, and $W(l, t)$, and the output of the network is the control signal u. The weight adjustment mechanism is based on the steepest descent method, namely

$$\dot{\mathbf{w}} = -\eta \left(\frac{\partial J}{\partial \mathbf{w}}\right)^T,$$

where $\mathbf{w}$ is the vector of the weights of the network and η is the learning rate. Now $\frac{\partial J}{\partial \mathbf{w}}$ is computed according to

$$\frac{\partial J}{\partial \mathbf{w}} = \frac{\partial J}{\partial e}\frac{\partial e}{\partial \mathbf{w}} + \frac{\partial J}{\partial \dot{e}}\frac{\partial \dot{e}}{\partial \mathbf{w}} + \frac{\partial J}{\partial W}\frac{\partial W}{\partial \mathbf{w}} =$$
$$e^T K_1 \frac{\partial e}{\partial \mathbf{w}} + \dot{e}^T K_2 \frac{\partial \dot{e}}{\partial \mathbf{w}} + W^T K_3 \frac{\partial W}{\partial \mathbf{w}}.$$

Since $e = y_r - \theta$, we get

$$\frac{\partial J}{\partial \mathbf{w}} = -e^T K_1 \frac{\partial \theta}{\partial \mathbf{w}} - \dot{e}^T K_2 \frac{\partial \dot{\theta}}{\partial \mathbf{w}} + W^T K_3 \frac{\partial W}{\partial \mathbf{w}},$$

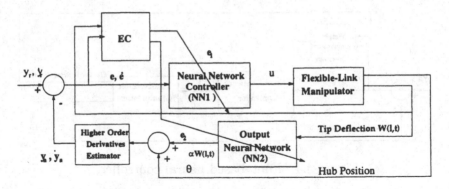

Figure 4.5. Structure of the neural network based controller using output redefinition. The block EC performs the linear combination of $e, \dot{e}$, and $W(l,t)$ for specific learning schemes to be used in NN1 and NN2.

and by using $\frac{\partial \theta}{\partial \mathbf{w}} = \frac{\partial \theta}{\partial u} \frac{\partial u}{\partial \mathbf{w}}$, $\frac{\partial \dot{\theta}}{\partial \mathbf{w}} = \frac{\partial \dot{\theta}}{\partial u} \frac{\partial u}{\partial \mathbf{w}}$, $\frac{\partial W}{\partial \mathbf{w}} = \frac{\partial W}{\partial u} \frac{\partial u}{\partial \mathbf{w}}$ and $u = \Phi(e, \dot{e}, W, \mathbf{w})$, we may write $\frac{\partial J}{\partial \mathbf{w}}$ as

$$\frac{\partial J}{\partial \mathbf{w}} = (-e^T K_1 \frac{\partial \theta}{\partial u} - \dot{e}^T K_2 \frac{\partial \dot{\theta}}{\partial u} + W^T K_3 \frac{\partial W}{\partial u}) \frac{\partial \Phi}{\partial \mathbf{w}},$$

where $\frac{\partial \Phi}{\partial \mathbf{w}}$ can be computed using the backpropagation method, and $\frac{\partial \theta}{\partial u}, \frac{\partial \dot{\theta}}{\partial u}$, and $\frac{\partial W}{\partial u}$ are computed as suggested in [104] by using the sign of the gradient instead of its real value for training the neural controller.

4.5.   Output Redefinition Through Online Learning (ORTOL)

In this section, the assumption of *a priori* knowledge about the linear model of the system is relaxed through online learning. The proposed control structure is shown in Figure 4.5.

In this structure, two neural networks are employed. The first neural network (NN1) is trained to function as a feedback controller and the second neural network (NN2) is trained to provide a proper output for feedback. In other words, NN1 is trained to produce a control action so that the error between the output defined by NN2 and the desired reference trajectory is minimized. The objective function that is used for training NN1 is written as $J_1 = \frac{1}{2}(e^T K_1 e + \dot{e}^T K_2 \dot{e} + W(l,t)^T K_3 W(l,t))$, where $e := y_r - y_a$, y_a is constructed by measuring θ and adding to the output of NN2, and $W(l,t)$ is introduced to reduce the vibrations of the flexible modes of the system in the output. The objective function J_1 is a weighted function of e, $\dot{e}$ and $W(l, t)$ where the corresponding weights are specified by K_1, K_2 and K_3. The inputs to the NN1 network are e, $\dot{e}$ and $W(l, t)$, and the output of the network is the control signal u. The weight adjustment mechanism is based on the steepest descent gradient method, namely

$$\dot{\mathbf{w}} = -\eta_1 \left(\frac{\partial J_1}{\partial \mathbf{w}} \right)^T,$$

$$\frac{\partial J_1}{\partial \mathbf{w}} = \frac{\partial J_1}{\partial e} \frac{\partial e}{\partial \mathbf{w}} + \frac{\partial J_1}{\partial \dot{e}} \frac{\partial \dot{e}}{\partial \mathbf{w}} + \frac{\partial J_1}{\partial W} \frac{\partial W}{\partial \mathbf{w}} =$$
$$e^T K_1 \frac{\partial e}{\partial \mathbf{w}} + \dot{e}^T K_2 \frac{\partial \dot{e}}{\partial \mathbf{w}} + W^T K_3 \frac{\partial W}{\partial \mathbf{w}}.$$

Using $e = y_r - y_a$, we get

$$\frac{\partial J_1}{\partial \mathbf{w}} = -e^T K_1 \frac{\partial y_a}{\partial \mathbf{w}} - \dot{e}^T K_2 \frac{\partial \dot{y}_a}{\partial \mathbf{w}} + W^T K_3 \frac{\partial W}{\partial \mathbf{w}}.$$

Now, using $\frac{\partial y_a}{\partial \mathbf{w}} = \frac{\partial y_a}{\partial u} \frac{\partial u}{\partial \mathbf{w}}$, $\frac{\partial \dot{y}_a}{\partial \mathbf{w}} = \frac{\partial \dot{y}_a}{\partial u} \frac{\partial u}{\partial \mathbf{w}}$, $\frac{\partial W}{\partial \mathbf{w}} = \frac{\partial W}{\partial u} \frac{\partial u}{\partial \mathbf{w}}$ and $u = \Phi(e, \dot{e}, W, \mathbf{w})$, we may write $\frac{\partial J_1}{\partial \mathbf{w}}$ as

$$\frac{\partial J_1}{\partial \mathbf{w}} = \left(-e^T K_1 \frac{\partial y_a}{\partial u} - \dot{e}^T K_2 \frac{\partial \dot{y}_a}{\partial u} + W^T K_3 \frac{\partial W}{\partial u} \right) \frac{\partial \Phi}{\partial \mathbf{w}},$$

where $\frac{\partial \Phi}{\partial \mathbf{w}}$ can be computed using the backpropagation method, and $\frac{\partial y_a}{\partial u}$, $\frac{\partial \dot{y}_a}{\partial u}$, and $\frac{\partial W}{\partial u}$ are computed as suggested in [104] by using the sign of the gradient instead of its real value for training the neural controller. An approximation to the gradient was also suggested in [105] as

$$\frac{\partial y}{\partial u} = \frac{y(u + \delta u) - y(u)}{\delta u}. \tag{4.7}$$

This approximate derivative can be determined by changing each input to the plant slightly at the operating point and measuring the changes with previous iterations. Using the sign of gradient, the error of differentiation can be avoided.

The objective of the NN2 network is to generate an output of the form $y_{ai} = \theta_i + \alpha_i \frac{W_i(l_i, t)}{l_i}$. Since θ_i can be measured and $W_i(l_i, t)$ can be computed from y_{ti} and θ_i, a neural network can be trained to obtain an appropriate estimate for α_i. Using a network whose weights are limited to the range $[-1, 1]$, an objective function that can be minimized is selected as $J_2 = \frac{1}{2}(e^T e)$. This leads to

$$\dot{\alpha}_i = -\eta_2 \frac{\partial J_2}{\partial \alpha_i} = -\eta_2 e_i \frac{\partial e_i}{\partial \alpha_i} = \eta_2 e_i \frac{\partial y_{ai}}{\partial \alpha_i} = \eta_2 e_i W_i(l_i, t)$$

The input to the network is $W(l, t)$ and the output vector elements are computed as $\alpha_i W_i(l_i, t)$. The new output that is used for feedback is now constructed as $y_{ai} = \theta_i + \alpha_i \frac{W_i(l_i, t)}{l_i}$.

4.6.   Simulation Results

4.6.1.   A Single Flexible–link Manipulator

In this section, simulation results for the proposed neural network controllers are presented. For comparing the performance of the new output with that of the RTP, two single flexible–link systems namely System I and II, with different flexibilities are considered in the simulations. The link parameters for System I are given in Table B.1 and those for System II are given in Table 2.1. Numerical models for the two systems are given in Appendix B. The first two frequencies for System I are at 16 and 100 rad/s while those for System II are at 3 and 19 rad/s.

Conventional PD Control

Figure 4.6 shows the step response of System II obtained using conventional PD control for $\alpha = 0.7$. As can be seen (Figure 4.6–c and 4.6–d) when Coulomb friction is included in the model, the PD control fails to give an acceptable response.

The IDML Scheme

The feedback–error learning methods were successfully tested in performing inverse dynamics control based on the new output. A three-layer neural network was used with 4 neurons in the input layer, 5 neurons in the hidden layer, and 1 neuron in the output layer. The inputs to the network are $\dot{\theta}, W, \dot{W}, \ddot{y}_a$. The hidden layer neurons have sigmoidal transfer functions and the output neuron uses a linear activation function. The conventional controller is given by (4.3). The following results were obtained with $K_2 = 1, K_1 = 2$ and $K_0 = 1$. Figure 4.7 shows the simulation results for System II obtained by using the new output ($\alpha = 0.7$) when no friction is included in the model and the payload mass is $M_l = .05\ Kg$, but is assumed to be unknown. As can be seen from Figure 4.7 the tip position tracking is considerably improved compared to that obtained using PD control (Figure 4.6). To show the robustness of the neural network controller when hub friction is included in the model, simulation results were obtained as shown in Figure 4.8. As can be seen, the neural network–based controller still yields excellent response.

The NRL Scheme

For implementing the NRL scheme, a three-layer neural network was used with 8 neurons in the input layer, 5 neurons in the hidden layer, and 1 neuron in the output layer. The inputs to the network are $\dot{\theta}, W, \dot{W}, e, \dot{e}, y_r, \dot{y}_r$, and $\ddot{y}_r$, where $e = y_r - y_a$. The hidden layer neurons have sigmoidal transfer functions and the output neuron uses a linear activation function. The conventional controller is given by (4.3). The following results are obtained with $K_2 = 1, K_1 = 2$ and $K_0 = 1$.

Figures 4.9 and 4.10 show the simulation results obtained using the RTP for System I and II, respectively. As can be seen from Figure 4.9, the RTP control works well for a less flexible system (System I) with a low speed desired trajectory namely, $sin(t)$. However, as Figure 4.10 reveals, for a more flexible system (System II), while RTP tracks the desired trajectory (top figure), there is significant tracking error in the tip position response (middle figure). Figure 4.11 shows the simulation results for System II obtained by using the new output ($\alpha = 0.7$) when no friction is included in the model and the payload mass is $M_l = .05\ Kg$, but is assumed to be unknown. In Figure 4.11, the results obtained by the conventional inverse dynamics control when all the nonlinearities are known and all the states are available are compared to the result obtained by using the new output with a neural network controller. As can be seen the neural network has successfully learned the inverse dynamics of the system. The robustness of the NRL scheme is also examined by including the hub friction in the model. Figure 4.12 demonstrate the results.

In Figure 4.13, the effects of the modification of the learning rule of the neural network for controlling System II are shown. This modification was done by adding $W(l, t)$ to the error signal to control the vibrations of the flexible modes of the system. Specifically when $\alpha = 0$, the hub response is very smooth whereas the response of the tip is quite oscillatory (top figure). Now by adding $W(l, t)$ to the error signal, the response of the tip and hub positions (shown in the bottom figure) reveals that the vibrations of the flexible modes are considerably reduced.

The Joint–based Control Scheme

Joint–based controller was implemented using a three–layer neural network with 20 hidden neurons 3 input neurons and 1 output neuron. The inputs to the network are e, $\dot{e}$, $W(l, t)$ where $e = y_r - \theta$. The activation function used for the input and hidden layers is the tan–sigmoid function and for the output layer is a linear function.

Figure 4.14 shows the simulation result for System II obtained using the joint–based control when no friction is included in the model and the payload mass is $M_l = .05 \ Kg$, but is assumed to be unknown. As can be seen from Figure 4.14-a while joint position tracks the reference trajectory, the tip position tracks the desired trajectory with small tracking errors (Figure 4.14-b).

The effects of modifying the learning rule for controlling System II are shown in Figure 4.15. This modification was done by adding $W(l, t)$ to the error signal to control the vibrations of the flexible modes of the system. As can be observed, when the objective function is not modified, even though that the joint response is very smooth, the response of the tip is quite oscillatory (Figure 4.15-a). Now by adding $W(l, t)$ to the error signal, the response of the tip and joint positions (Figure 4.15-b) reveals that the vibrations of the flexible modes are considerably reduced.

To show the robustness of the neural network controller when hub friction is included in the model, simulation results are shown in Figure 4.16. As can be seen, the actual tip response does not change when Coulomb friction is included in the model.

The ORTOL Scheme

Simulations were performed using the fourth scheme in which the *a priori* knowledge about the system dynamics has been relaxed through online learning. A three–layer neural network was employed for NN1 with 3 input neurons, 20 hidden neurons and 1 output neuron. The activation function used for the input and hidden layers is the tan–sigmoid function and for the output layer is a linear function. Figure 4.17 shows the simulation results for System II with a $sin(t)$ reference trajectory. Figure 4.17–a shows the response of the output defined by NN2. This clearly shows exact tracking of the desired trajectory. Figure 4.17–b is the response of the actual tip showing a small steady–state tracking error. Figure 4.17–c shows the evolution of α. These results demonstrate that very good response is obtained even when no *a priori* knowledge about the system dynamics is incorporated in designing the neural network–based controller.

4.6.2. A Two–link Planar Manipulator

Simulation results for a two–link planar manipulator are presented in this section. The manipulator consists of one rigid arm (first link) and one flexible arm (second link) with the following numerical data [43]

$$l_1 = 20cm, l_2 = 60cm, A_1 = 3.14cm \times 1.3cm, A_2 = 5cm \times 0.9mm,$$
$$\rho_1 = 2700kg/m^3 (6061 Aluminum), \rho_2 = 7981 (Stainless Steel),$$
$$M_1 = 1kg, M_l = 0.251kg, m1 = 0.236kg, m_2 = 0.216kg,$$
$$E = 194 \times 10^9 N/m^2, J_1 = 0.11 \times 10^{-3} kgm^2,$$
$$J_l = 0.11 \times 10^{-4}, J_h = 3.8 \times 10^{-5}$$

where l_1 and l_2 are link lengths, A_1 and A_2 are cross sectional areas, E and ρ are modulus of elasticity and mass density, J_h is the hub inertia and M_1, M_l, J_1 and J_l are masses and mass moment of inertia at the end points of the two links. The first two natural frequencies of the second link are 5.6 and $27.6Hz$

First, based on the procedure given in Chapter 3 the value of α^* is found to be $\alpha^* = [1 \ 0.6]^T$. Consequently, a value of $\alpha_2 = 0.5$ is used in the simulations. Figures 4.18–a to 4.18–d show the system responses to a $sin(t)$ reference trajectory for both links. These responses are obtained by using a conventional PD controller for the new output ($\alpha_2 = 0.5$). As can be seen from Figures 4.18–a to 4.18–d, considerable amounts of tracking errors are present in the responses of $\theta_1, y_{t1}, \theta_2$ and y_{a2}. As a comparison, the responses of the system to the same reference trajectory which were obtained using neural network controllers are shown in Figure 4.19 (IDML scheme), Figure 4.20 (NRL scheme), Figure 4.21 (joint–based control scheme), and Figure 4.22 (ORTOL scheme).

The structures of the neural networks employed for these simulations are similar to those of neural networks explained in Section 4.6.1. The differences are that 2 neurons are used in the output layer of each network for all proposed schemes and the number of hidden neurons has been increased to 10 for the IDML and NRL schemes. This is due to the increased complexity in the dynamics of a two–link manipulator as compared to those of a single–link manipulator.

These figures demonstrate significant improvement in the responses of the system $(\theta_1, y_{t1}, \theta_2$ and $y_{a2})$. As Figures 4.19 and 4.20 display, the IDML and the NRL schemes yield similar results for this system since both schemes use the same learning rule, that is feedback–error learning. As compared to the results obtained using the joint–based control and the ORTOL schemes (see Figures 4.21 and 4.22), more accurate result can be obtained using the IDML and the NRL schemes. This is due to the fact that the IDML and the NRL schemes use some *a priori* knowledge about the system dynamics (a linear model of the system).

Having no *a priori* knowledge about the system dynamics also leads to an increase in the sizes of the neural networks. For instance, the IDML and the NRL schemes are able to obtain good tracking performance with 10 hidden neurons while good results for the joint–based control and the ORTOL schemes are obtained by using 20 hidden neurons.

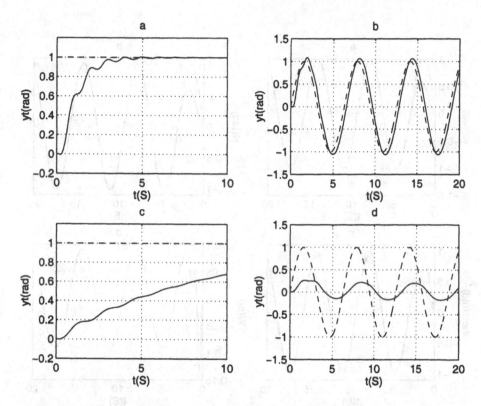

Figure 4.6. Actual tip responses for step and $sin(t)$ reference trajectory for System II using PD control of the new output with and without friction at the hub: (a) step response without friction, (b) response without friction for $sin(t)$, (c) step response with friction, (d) response with friction for $sin(t)$. (dashed lines correspond to the desired trajectories).

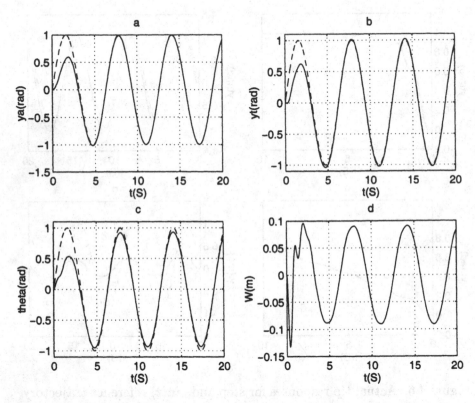

Figure 4.7. Output responses for $sin(t)$ reference trajectory for System II using the IDML neural network controller: (a) Redefined output y_a, (b) Actual tip position y_t, (c) Joint position θ and (d) Tip deflection $W(l,t)$. (dashed lines correspond to the desired trajectories).

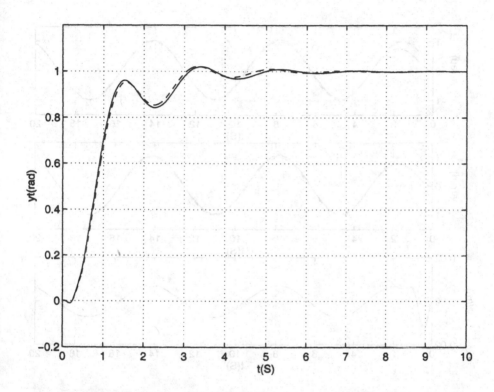

Figure 4.8. Actual tip responses to step input for System II using the IDML neural network controller; (dashed line corresponds to model with Coulomb friction at the hub).

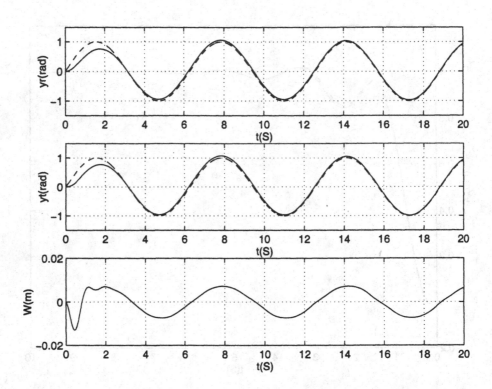

Figure 4.9. Output responses for $sin(t)$ reference trajectory for System I using the RTP output: RTP (top), actual tip position (middle), total deflection (bottom) (dashed line corresponds to the desired trajectory).

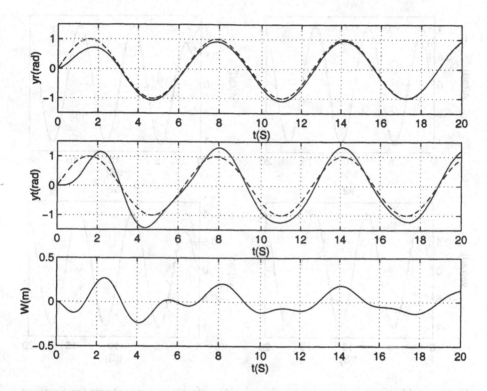

Figure 4.10. Output responses for $sin(t)$ reference trajectory for System II using the RTP output: RTP (top), actual tip position (middle), total deflection (bottom) (dashed line corresponds to the desired trajectory).

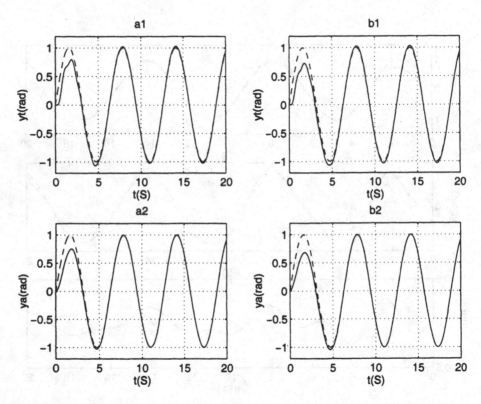

Figure 4.11. Output responses for $sin(t)$ reference trajectory for System II using the new output y_a, $\alpha = 0.7$: (a1) actual tip position (NN), (a2) redefined output (NN), (b1) actual tip position (inverse dynamics), (b2) redefined output (inverse dynamics) (dashed line corresponds to the desired trajectory).

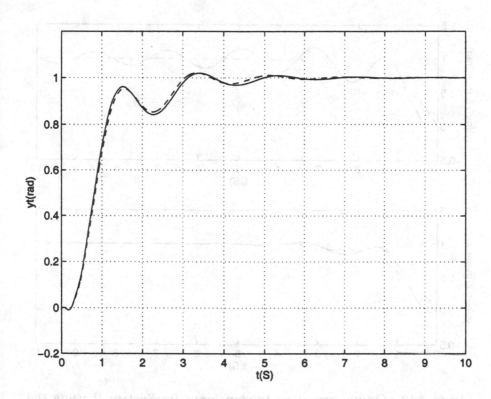

Figure 4.12. Actual tip responses to step input for System II using the NRL neural network controller; (dashed line corresponds to the model with Coulomb friction at the hub).

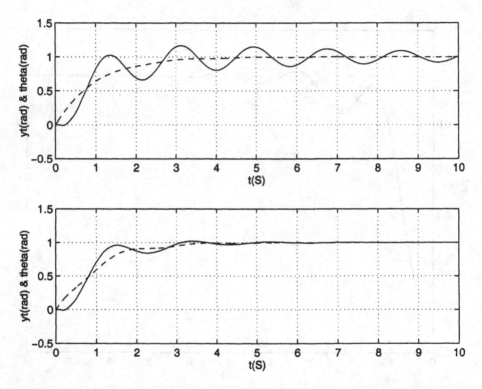

Figure 4.13. Output responses to step input for System II using the NRL neural network controller: Actual tip position (solid lines), hub position (dashed lines) (bottom figure corresponds to the modified version of the neural network controller).

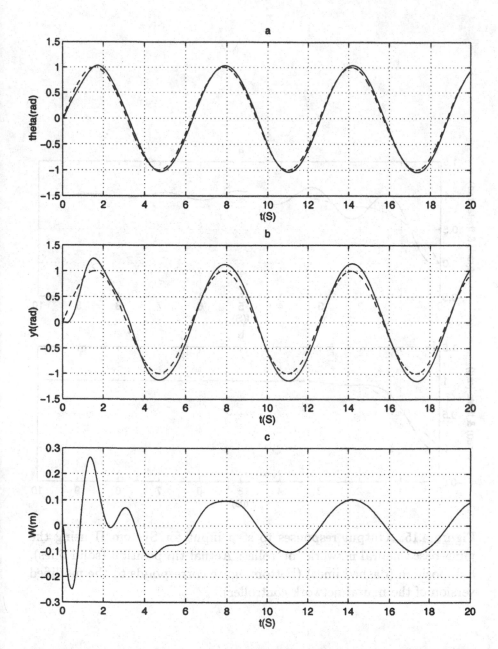

Figure 4.14. Output responses for *sin(t)* reference trajectory for System II using the joint–based neural network controller: (a) Joint position, (b) actual tip position (middle) and (c) total deflection (bottom) (dashed line corresponds to the desired trajectory).

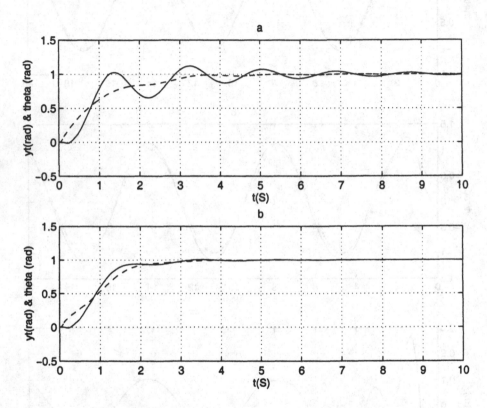

Figure 4.15. Output responses to step input for System II using the joint–based neural network controller: Actual tip position (solid lines), hub position (dashed lines) (bottom figure corresponds to the modified version of the neural network controller).

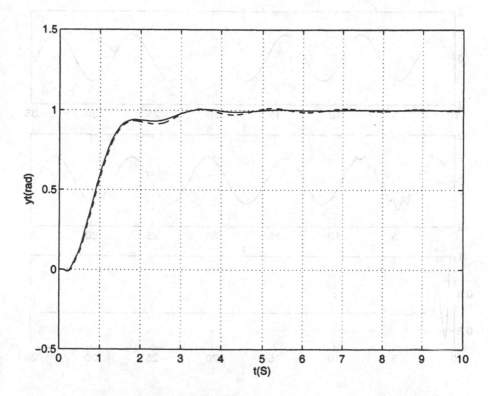

Figure 4.16. Actual tip responses to step input for System II using the joint–based neural network controller; (dashed line corresponds to the model with Coulomb friction at the hub).

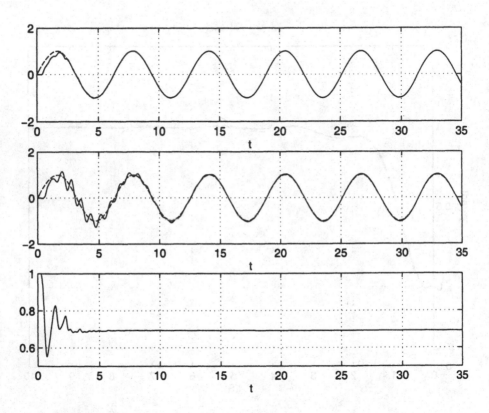

Figure 4.17. Output responses for $sin(t)$ reference trajectory for System II using the ORTOL neural network controller: (a) Redefined output y_a, (b) actual tip position y_t and (c) evolution of α (dashed line corresponds to the desired trajectory).

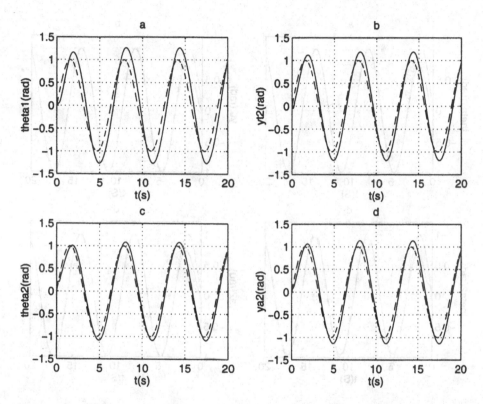

Figure 4.18. System responses for $sin(t)$ reference trajectory using a PD controller for the redefined output y_a ($\alpha = 0.5$): (a) Joint position (first link), (b) actual tip position (second link), (c) Joint position (second link), and (d) redefined output (second link) (dashed lines correspond to the desired trajectories).

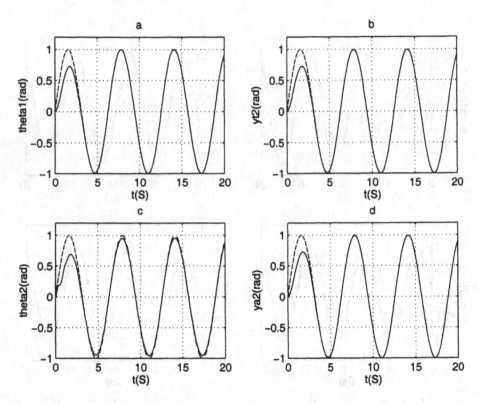

Figure 4.19. System responses for $sin(t)$ reference trajectory using the IDML neural network controller for the redefined output y_a ($\alpha = 0.5$): (a) Joint position (first link), (b) actual tip position (second link), (c) Joint position (second link), and (d) redefined output (second link) (dashed lines correspond to the desired trajectories).

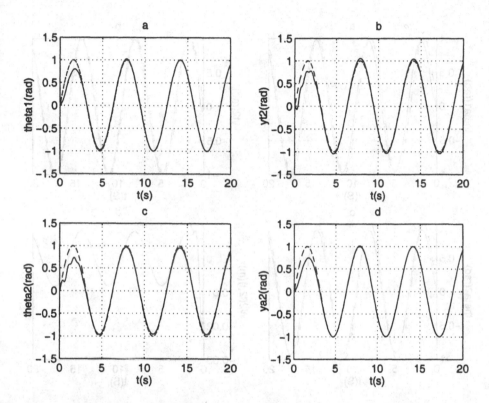

Figure 4.20. System responses for $sin(t)$ reference trajectory using the NRL neural network controller for the redefined output y_a ($\alpha = 0.5$): (a) Joint position (first link), (b) actual tip position (second link), (c) Joint position (second link), and (d) redefined output (second link) (dashed lines correspond to the desired trajectories).

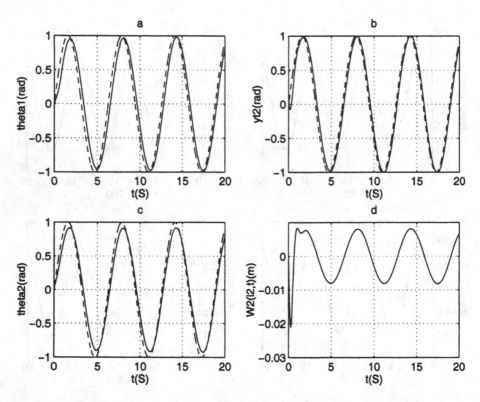

Figure 4.21. System responses for $sin(t)$ reference trajectory using the joint–based neural network controller: (a) Joint position (first link), (b) actual tip position (second link), (c) Joint position (second link), and (d) Total tip deflection (second link) (dashed lines correspond to the desired trajectories).

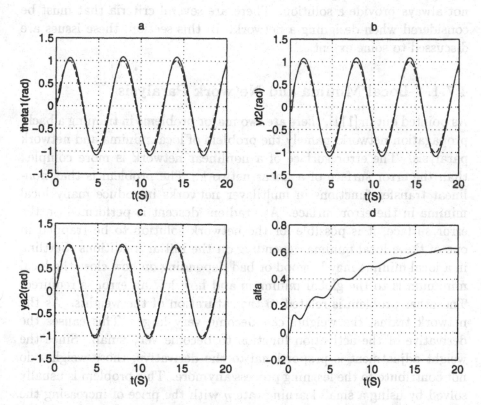

Figure 4.22. System responses for $sin(t)$ reference trajectory using the ORTOL neural network controller: (a) Joint position (first link), (b) actual tip position (second link), (c) redefind output (second link), and (d) evolution of α (second link) (dashed lines correspond to the desired trajectories).

4.7. Limitations and Design Issues

Multilayer neural networks are capable of performing almost any linear or nonlinear computation and can approximate any reasonable function arbitrarily well. However, while the trained network is theoretically capable of performing a mapping, the backpropagation algorithm may not always provide a solution. There are several criteria that must be considered when designing a network. In this section, these issues are discussed to some extent.

4.7.1. Local Minima and Network Paralysis

As pointed out in [111], there are two major problems in training a back-propagation network, namely the problem of local minima and network paralysis. The error surface of a nonlinear network is more complex than the error surface of a linear network. The problem is that non-linear transfer functions in multilayer networks introduce many local minima in the error surface. As gradient descent is performed on the error surface, it is possible for the network solution to be trapped in one of these local minima depending on the initial conditions. Settling in a local minima may be good or bad depending on how close the local minimum is to the global minimum and how low an error is required. The other problem is related to the saturation of the weights. As the network trains, the weights can become very large. This causes the derivative of the activation function to become very small. Since the weight adjustment is proportional to the derivative, those weights do not contribute to the learning process anymore. This problem is usually solved by using a small learning rate η with the price of increasing the learning time.

4.7.2. Learning Rate

The effectiveness and convergence of the error backpropagation algorithm depend significantly on the value of the learning rate η. In general however, the optimum value of η depends on the problem being solved and there is no single learning rate value suitable for different problems. This problem seems to be common for all gradient based optimization schemes.

When flat minima yield small gradient values, a large value of η will result in more rapid convergence. However, for problems with steep and narrow minima, a small value of η must be chosen to avoid oscillation around the minima. This leads to the conclusion that η should indeed be chosen experimentally for each problem. One should also remember that only small learning rates guarantee a true gradient descent. The price of this guarantee is an increase in the total number of learning steps. Although the choice of the learning rate depends strongly on the class of the learning problem and on the network architecture, values ranging from 10^{-3} to 10 have been reported [112].

For large learning rates, the learning speed can be drastically increased; however, the learning may not be exact, with tendencies for fluctuations or it may never stabilize at any minima. One solution is to choose a large value of η in the early stage of learning when the error is large and decrease it as learning evolves and the error becomes smaller. Even though simple gradient descent can be efficient, there are some methods for improving the rate of convergence.

Momentum Method

The purpose of the momentum method is to accelerate the convergence of the error backpropagation algorithm. This method involves adding a fraction of the most recent weight adjustment to the current weight adjustment. For the discrete-time neural network this is done by setting

$$\Delta \mathbf{w}_k = -\eta \frac{\partial J_k}{\partial \mathbf{w}_k} + \alpha \Delta \mathbf{w}_{k-1}$$

where J_k is the objective function to be minimized, $\Delta \mathbf{w}_k$ and $\Delta \mathbf{w}_{k-1}$ represent the current and previous weight adjustment respectively and α is a positive momentum constant. The second term, indicating a scaled most recent adjustment of weights, is called the momentum term which tends to keep the weight changes going in the same direction.

Adaptive Learning Rate

In this method, the ratio of the current output error to the previous one is checked at every step to determine whether the training shows a convergent or divergent trend. Based on this, the current learning rate is increased or decreased by specified factors. This method increases the learning rate at times to speed up the training only to the extent that the network can learn without large error increases [113]. When a larger learning rate can result in stable learning, the learning rate is increased. When the learning rate is too high to guarantee a decrease in error, it gets reduced automatically until stable learning resumes. For our problem, we chose a fixed value of η for the first three learning schemes, but for the fourth learning scheme, an adaptive learning rate has been used for both neural networks (output neural network NN2, and neural controller NN1).

4.7.3. Slope of the Activation Function

The neuron's continuous activation function is characterized by its steepness. Also, the derivative of the activation function in the backpropagation algorithm [112] serves as a multiplying factor in building components of the error signal. Thus both the choice and the slope of the activation function strongly affects the speed of the network learning. A typical activation function can be written as

$$f(x) = \frac{2}{1 + e^{-ax}} - 1, \quad a > 0$$

and its derivative can be computed as

$$f'(x) = \frac{2ae^{-ax}}{(1 + e^{-ax})^2}$$

and it reaches a maximum value of $\frac{1}{2}a$ at $x = 0$. Since weights are adjusted in proportion to the value of $f'(x)$, the weights that are connected to units responding in their mid–range are changed the most. The weights of the uncommitted neurons are thus affected more strongly than of those neurons that are already turned on or turned off. The other feature of this derivative is that for a fixed learning rate all adjustments of weights are in proportion to the steepness coefficient a. This particular observation leads to the conclusion that using an activation function with large a may yield results similar to those for a large learning rate η. It thus seems advisable to keep a at a fixed value and to control the learning speed using only the learning rate η rather than controlling both η and a. In our experiments, we fixed the slope of the activation function as suggested in the MATLAB Neural Network Toolbox [113] in all cases ($a = 2$).

4.7.4. Initial Weights Selection

The choice of initial weights and biases plays a significant role in the training of the network. Typically, the weights of the network are initialized at small random values. The initialization strongly affects the final solution. If all weights start with equal values, and if the solution requires unequal weights, the network may not train properly. Unless the network is disturbed by random factors, the internal representation may continuously result in symmetric weights. Small initial values, however, make the speed of convergence very slow. In [114], a method of selecting initial weights and biases is proposed. This method of initialization is formulated based on the range of inputs and outputs. The random initial weights between $[-.5, .5]$ and zero biases are used for all of the neural network structures mentioned in the previous sections. For the experiment, however, we added a small bias to the network.

4.8.   Conclusions

In this chapter, four neural network–based control schemes were proposed for tip position tracking of a flexible manipulator. The first two schemes were developed by using a modified version of the "feedback–error learning" approach to learn the inverse dynamics of the flexible manipulator. Both schemes assume some *a priori* knowledge of the linear model of the system. This assumption was relaxed in the third and fourth schemes. In the third scheme, the controller was designed based on tracking the hub position while controlling the elastic deflection at the tip. The fourth scheme employed two neural networks, one of the neural networks defines an appropriate output for feedback and the other neural network acts as an inverse dynamics controller. Simulation results for two single flexible–link manipulators and a two–link manipulator were presented to illustrate the advantages and improved performance of the proposed tip position tracking controllers over the conventional PD–type controllers.

Chapter 5

Experimental Results

This chapter presents some experimental results obtained for a single flexible–link manipulator test-bed [91] constructed in our Robotics and Control Laboratory. In Section 5.1, the experimental test–bed will be described. Section 5.2 discusses some implementation issues and in Section 5.3 some experimental results will be presented. Finally, in Section 5.4, the modified schemes are presented.

5.1. The Test-bed

The experimental system [91] consists of a highly flexible link whose parameters are shown in Table 2.1. A schematic of the experimental test-bed is shown in Figure 5.1. The beam consists of a central stainless steel tube with annular surface corrugations. Aluminum blocks are bolted to the tube and two thin parallel spring steel strips slide within slots cut into the blocks. A high performance drive was assembled consisting of a pulse-width-modulated amplifier that operates in current feedback mode, a DC servo motor with an optical encoder and a harmonic drive speed reducer. An infrared emitting diode is used to sense the position of the tip. The detector consists of a UDT camera consisting of a lens and an infrared- sensitive planar diode, and is mounted at the link's hub. The digital controller consists of a Spectrum C30 system card, based on the Texas Instruments TMS320C30 digital signal processing chip that operates from a 33.3 MHz clock and achieves a performance of 16.7 million instruction per second. Two channels of 16 bit A/D and D/A are also provided. An interface system was designed and built to connect the Spectrum card to the current amplifier, infrared

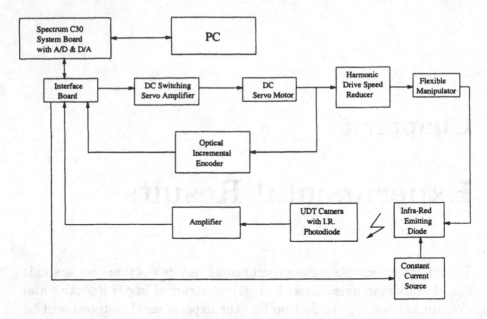

Figure 5.1. Block diagram of the experimental test bed.

emitting diode, optical encoder and infrared detector. The maximum torque range generated by the motor is ± 0.705 $N.m$. The speed reducer amplifies the motor torque by a factor of 50 and yields an output torque range of 35.25 $N.m$. The maximum tip deflection of $= \pm 0.25$ m can be measured using the infrared emitting diode, UDT camera and UDT amplifier. More details about the test-bed are given in Appendix C.

5.2.   Implementation Issues

5.2.1.   Selection of the Output

In earlier experiments on this test-bed [91], the author designed a linear controller based on transmission zero assignment to control the flexible-link system. In [91], the model of the manipulator was derived and except for the hub friction and viscous damping, good agreement between the experiment and the model was reported. The required parameters for the linear model of the manipulator have been taken from [91]. First, based on the reported values for the viscous damping parameters, i.e., $f_1 = 0.4$ and $f_2 = 4$, the value of α^* is found to be $\alpha^* = 0.75$. However, experimental results show that controlling the system using this output leads to *instability*. The source of this problem is the uncertainty in the coefficients f_1 and f_2. As the analysis in Appendix A shows a value of $\alpha^* = 0.6$ is robust to variations in f_1 and f_2 coefficients. However, in controlling the system using this output, the flexible modes of the system actually vibrate with such high amplitudes that in some situations the tip deflections exceed the feasible range of the sensor measurements. Consequently, a value of $\alpha = 0.48$ is used to ensure that the deflections in the arm are within the range of the sensor.

5.2.2. Estimation of the Higher-order Derivatives of the Output

As stated earlier, with the available sensors, we can only measure the joint position and the tip deflection while in our proposed control scheme we need first and second–order derivatives of the output. Towards this end, an observer has been designed to estimate the higher–order derivatives of the output y_a according to [92]:

$$\dot{\xi}_1 = \xi_2 + \frac{l_{n-1}}{\epsilon}(y_a - \xi_1),$$

$$\vdots \tag{5.1}$$

$$\dot{\xi}_{n-1} = \xi_n + \frac{l_1}{\epsilon^{n-1}}(y_a - \xi_1),$$

$$\dot{\xi}_n = \frac{l_0}{\epsilon^n}(y_a - \xi_1),$$

where the $l_i, i = 1, \ldots, n-1$ are chosen such that the polynomial $H(s) = s^n + l_{n-1}s^{n-1} + \ldots + l_1 s + l_0$ is Hurwitz, and ϵ is a sufficiently small positive number. The states $\xi_1, \ldots, \xi_n$ asymptotically estimate y_a and its higher order derivatives up to order $n - 1$. For details refer to [92].

5.2.3. Neural Network Structure

The proposed controllers have been implemented on a spectrum TMS320C30 real–time system board. Because of the discrete-time nature of the controller, continuous-time neural networks described in Chapter 4 were converted to discrete-time neural networks. The weights of this neural network are updated as $\Delta \mathbf{w}_k = -\eta \frac{\partial J(\mathbf{w})}{\partial \mathbf{w}}$. The conventional controller used is $u_c = K_2(\ddot{y}_r - \xi_3) + K_1(\dot{y}_r - \xi_2) + K_0(y_r - y_a)$, where ξ_2 and ξ_3 are obtained from (5.1). Two sampling frequencies were used , i.e. $200 \ Hz$ and $500 \ Hz$. These values were selected to satisfy the time requirements for computing the feedforward stage as well as back propagation stage of the neural networks. The following results for the neural networks were obtained with $f = 200\ Hz$. The results for the PD controllers were obtained with $f = 500\ Hz$. Note that the $200\ Hz$ sampling frequency for the PD controller leads to instability.

5.3.   Discussion of Results

In this section, experimental results for different control strategies are presented.

Conventional PD Control

In the first experiment using the new output y_a ($\alpha = 0.48$), a PD control with relatively large gains ($K_p = 100$, $K_v = 100$) was implemented. The responses of the actual tip and the new output are shown in Figure 5.2. As can be seen, there is a considerable amount of steady state–error in tracking tip position as well as the new output y_a. The errors are mainly caused by the presence of friction and stiction at the hub whose amplitudes vary with the hub position. The only way that PD control can overcome the effects of the friction and stiction at the hub is to increase its gains. Increasing the velocity gain K_v leads to *instability* of the system. As the PD gain K_p is increased, smaller steady–state errors are obtained however at the expense of a considerably oscillatory transient response at the tip. The results are shown in Figure 5.3. Note that the use of high gains may also lead to instability of the closed–loop system caused by saturation of the amplifier and large amplitude vibrations of the flexible modes. Using the gains $K_p = 200$ and $K_v = 100$, the system becomes **unstable** for a step input that is greater than $0.2\ rad$.

Next, M_l was increased to 850 g from 30 g and the responses of the system to a 0.2 rad step input are displayed in Figure 5.4. As can be observed, the responses exhibit less oscillation but greater steady–state error as compared to the $M_l = 30\ g$ case (see Figure 5.3).

The IDML Scheme

The IDML scheme was implemented based on the new output y_a ($\alpha = 0.48$) and by using a three-layer neural network with 4 neurons in the input layer, 5 neurons in the hidden layer, and 1 neuron in the output layer. The inputs to the network are $\dot{\theta}, W, \dot{W}, \ddot{y}_a$.   The hidden layer neurons have sigmoidal transfer functions and the output neuron uses a linear activation function. The conventional controller is given by equation (4.3). The responses of the actual tip and the new output to a 0.1 rad step input for $M_l = 30$ g case are shown in Figure 5.5, which clearly illustrates the improvements in the tracking performance as compared to Figures 5.2 and 5.3.

Next, $W(l,t)$ was incorporated to modify the learning rule of the neural network as in Section 4.3 to further improve the transient response of the system. The responses of the system are shown in Figure 5.6. This figure shows the improvement of the responses by using this modification. Figure 5.7 shows the responses of the actual tip and the new output to a 0.3 $sin(t)$ reference trajectory. It follows that the tip also tracks the desired trajectory with a small tracking error. Note that the PD control with the gains $K_p = 200$ and $K_v = 100$ leads to an **unstable** response for this desired trajectory.

Finally the robustness of the IDML scheme was examined by increasing the payload mass M_l from 30 g to 850 g. The responses of the system are shown in Figure 5.8. As Figures 5.6 and 5.8 demonstrate, the neural network controller is robust to the payload variation.

The NRL Scheme

By using the new output y_a ($\alpha = .48$), the NRL scheme was employed to control the system. A three-layer neural network was used with 8 neurons in the input layer, 5 neurons in the hidden layer, and 1 neuron in the output layer. The inputs to the network are $\dot{\theta}, W, \dot{W}, e, \dot{e}, y_r, \dot{y}_r$, and $\ddot{y}_r$, where $e = y_r - y_a$. The hidden layer neurons have sigmoidal activation functions and the output neuron uses a linear activation function. The conventional controller used is given by (4.3). The responses of the actual tip and the new output to a 0.1 rad step input are shown in Figure 5.9 which shows an improvement in tracking performance as compared to Figures 5.3 and 5.5.

The effect of the deflection control on the performance of the NRL method was also investigated. For this purpose, two experiments were performed with two values of α for the output namely, $\alpha = 0.48$ and $\alpha = 0$. The results for the tip position to 0.6 rad step input for both cases are shown in Figure 5.10. The 0.6 rad step input was chosen to demonstrate that the dynamic range of the neural network controller is much larger than that of the PD controller. Specifically, the PD controller yields an **unstable** closed-loop system. For comparison, the tip and hub responses for the $\alpha = 0$ case without any modification to the objective function are also shown. As can be seen, even for the $\alpha = 0$ case, the tip response is significantly better than that of a no modification case in the sense that the vibrations of the flexible modes are damped out very quickly.

The responses of the system to a $0.3\ sin(t)$ reference trajectory are shown in Figure 5.11. It follows that the tip also tracks the desired trajectory with a small tracking error. The robustness of the NRL scheme to the payload variation was examined by changing M_l to 850 g. The responses of the system in this case are depicted in Figure 5.12. As can be seen, NRL scheme is also robust to the payload variations.

The Joint–based Control Scheme

The joint–based control scheme was implemented by using a three-layer neural network with 3 neurons in the input layer, 5 neurons in the hidden layer and 1 neuron in the output layer. The inputs to the network are $e, \dot{e}$ and W, where $e = y_r - \theta$. The hidden layer neurons have sigmoidal activation functions and the output neuron uses a linear activation function.

The responses of the system to a 0.2 rad step input are shown in Figure 5.13, when in the cost function of the network $W(l,t)$ is not included (Figure 5.13–a and 5.13–b) and when it is included (Figure 5.13–c and 5.13–d). As can be seen, the tip response is significantly improved by adding $W(l,t)$ to the cost function of the neural network.

Figure 5.14 shows the responses of the system to a $0.2 + 0.1sin(t)$ reference trajectory. It can be observed that the tip also tracks the desired trajectory with small tracking errors. Next, M_l was increased to 850 g and the performance of the joint–based control scheme was evaluated under this change. The responses of the system are displayed in Figure 5.15. These figures demonstrate the robustness of the joint–based control scheme to the payload variation.

The ORTOL Scheme

The final set of experiments was performed using the fourth scheme in which requirement for the *a priori* knowledge about the system dynamics has been relaxed through online learning. A three–layer neural network was employed for NN1 with 3 input neurons, 5 hidden neurons and 1 output neuron. The activation function used for the input and hidden layers is the tan–sigmoid function and for the output layer is a linear function. Figure 5.16 shows the system responses to a 0.2 rad step input for $M_l = 30$ g. In Figure 5.16–d, evolution of α is shown. The responses of the system for $M_l = 850$ g are shown in Figure 5.17. Finally, in Figure 5.18, the responses of the system to a $0.2 + 0.1sin(t)$ are shown.

From these figure, we can conclude that good tracking performance can be obtained experimentally even when no *a priori* knowledge about the system dynamics is assumed.

As the figures demonstrate, all four proposed neural network schemes perform better than their conventional PD controller counterparts. This is due to the fact that to overcome the effects of friction using PD control, the PD gains have to be increased which considerably affects the transient response, dynamic range, and robustness of the closed–loop system. Among the neural network controllers, however, the IDML and the NRL schemes yield similar results (Figures 5.5 to 5.8 for IDML and 5.9 to 5.12 for NRL) since both schemes assume some *a priori* knowledge about the linear model of the system and use the same learning rule, i.e., feedback–error learning. However, the responses obtained by using the NRL scheme are more accurate and smoother than those obtained by using the IDML scheme. The reason is that in the NRL scheme, the reference trajectories (i.e. position, velocity, and acceleration) are fed to the neural network. This generates the feedforward term which makes the transient response smoother.

When no *a priori* knowledge about the system dynamics is assumed, smooth transient response is not guaranteed. For instance, the step responses of the system with $M_l = 850$ g obtained by using the ORTOL scheme (Figure 5.17) exhibit more oscillations than those obtained by using the IDML scheme (Figure 5.8) or the NRL scheme (Figure 5.12). Furthermore, looking at the responses of the system to $sin(t)$ reference trajectories (Figures 5.7, 5.11, 5.14, 5.18) reveals that the second flexible mode of the system (19 rad/s) vibrates with higher magnitude for the joint–based controller (Figure 5.14) and the ORTOL scheme (Figure 5.18) as compared to the IDML scheme (Figure 5.7) and the NRL scheme (Figure 5.11).

Table 5.1 compares the performance of the conventional PD controller to those of the neural network–based controllers. In this table, M_l is the payload mass, DC represents deflection control, T_s is the settling time, ESS is the steady–state error, PO is the percentage over-

Table 5.1. Summary of the results

Scheme	$M_l(g)$		$T_s(S)$	ESS (%)	PO (%)	PU (%)
PD	30	$K_p = 100,\ K_v = 100$	–	33	–	–
	30	$K_p = 200,\ K_v = 100$	6.69	0	34.5	48.2
	850	$K_p = 200,\ K_v = 100$	3.93	10	15	7.5
IDML	30	Without DC	5.56	0	8.2	12
	30	With DC	2.61	0	3.7	–
	850	With DC	2.96	0	2.0	–
NRL	30	Without DC	3.92	0	6.3	2.5
	30	With DC	3.00	0	2.1	–
	850	With DC	2.95	0	1.9	–
JBC	30	Without DC	5.46	0	12.6	36.2
	30	With DC	3.48	0	3.6	–
	850	With DC	3.28	0	3.1	–
ORTOL	30	With DC	4.55	0	2.3	–
	850	With DC	3.45	0	13	7.1

shoot, and PU is the percentage undershoot. The results are obtained for a 0.1 *rad* step input. The results given in this table lead to the conclusion that in general the neural network controllers are more robust to payload variations than the PD controller. The other conclusion that can be drawn is that deflection control significantly improves the responses in the sense that in general it removes the overshoot and undershoot from the responses and reduces the settling time.

To show the robustness of the neural network controllers to disturbances, another experiment was performed and the results are shown in Figure 5.19. First, a 0.3 *rad* step input was applied and the NRL neural controller was employed for controlling the system. After, the system reached its steady–state value, disturbances were applied as unexpected tip deflections (see Figure 5.19–d). As these figures show, the neural network controller can maintain a stable closed–loop system even when the magnitude of the disturbance reaches 0.3 *m*.

Finally, the last experiment attempts to verify the claim stated in Chapter 3, namely that *the region of minimum phase behavior increases as the payload mass increases.* As stated in Section 5.2.1, for $M_l = 30\ g$ it is not possible to control the system by using the new output y_a, $\alpha = 0.6$. Figure 5.20 shows the responses of the system to a $0.2 + 0.1sin(t)$ reference trajectory for $M_l = 850\ g$ obtained by using the NRL scheme ($\alpha = 0.65$). It can be observed that a stable closed–loop system is obtained with a small tracking error for the tip position. This validates and confirms the statement that *the value of α^* increases as the payload mass M_l increases.*

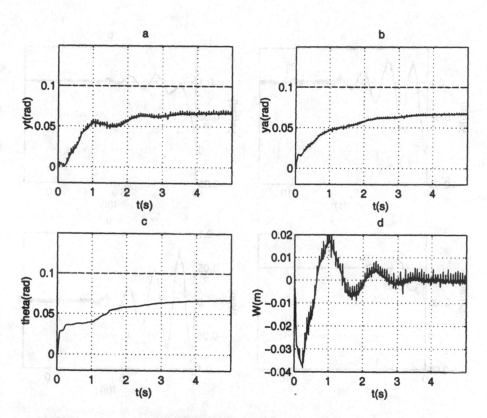

Figure 5.2. System responses to a 0.1 *rad* step input using a PD controller ($\alpha = .48$): (a)-actual tip position, (b)-redefined output (c)-hub position, and (d)-tip deflection (dashed lines correspond to the reference trajectories).

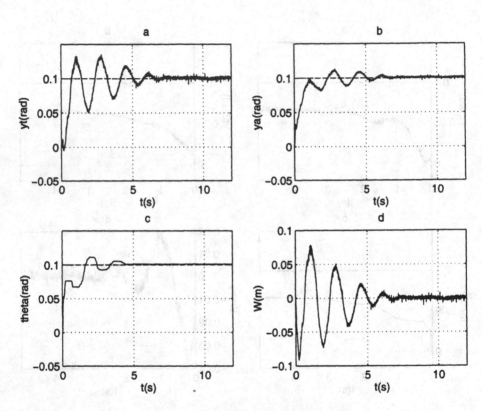

Figure 5.3. System responses to a 0.1 *rad* step input using a PD controller with higher K_p and K_v gains ($\alpha = .48$): (a)-actual tip position, (b)-redefined output (c)-hub position, and (d)-tip deflection (dashed lines correspond to the reference trajectories).

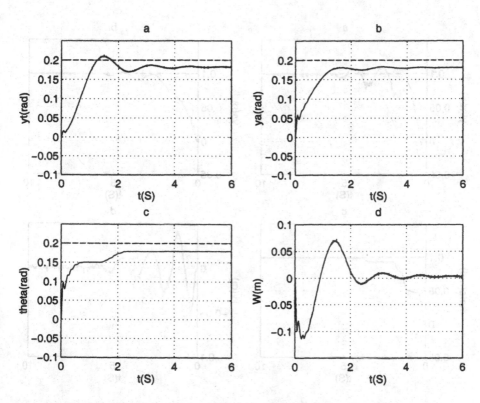

Figure 5.4. System responses to a 0.2 *rad* step input using a PD controller with $M_l = 850\ g$ ($\alpha = .48$): (a)-actual tip position, (b)-redefined output (c)-hub position, and (d)-tip deflection (dashed lines correspond to the reference trajectories).

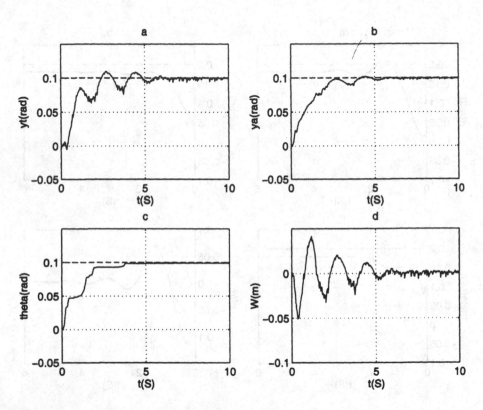

Figure 5.5. System responses to a 0.1 *rad* step input using the IDML neural network controller ($\alpha = 0.48$): (a) actual tip position, (b) redefined output (c) hub position, and (d) tip deflection (dashed lines correspond to the reference trajectories).

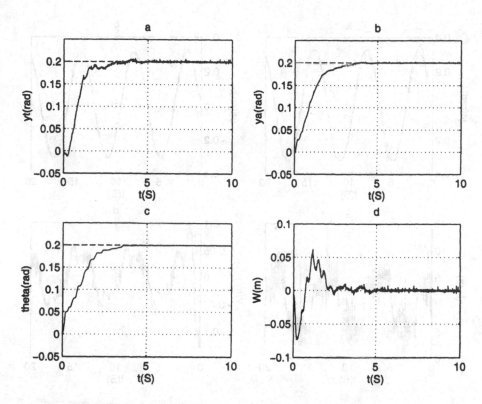

Figure 5.6. System responses to a 0.2 *rad* step input using the IDML neural network controller ($\alpha = 0.48$ with modified learning rule): (a) actual tip position, (b) redefined output (c) hub position, and (d) tip deflection (dashed lines correspond to the reference trajectories).

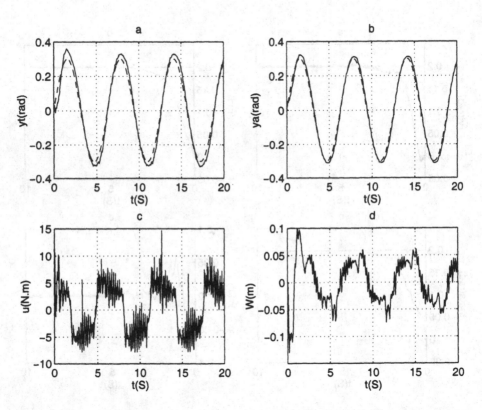

Figure 5.7. System responses to a 0.3 $sin(t)$ input using the IDML neural network controller ($\alpha = 0.48$): (a) actual tip position, (b) re-defined output (c) control torque, and (d) tip deflection (dashed lines correspond to the desired trajectories).

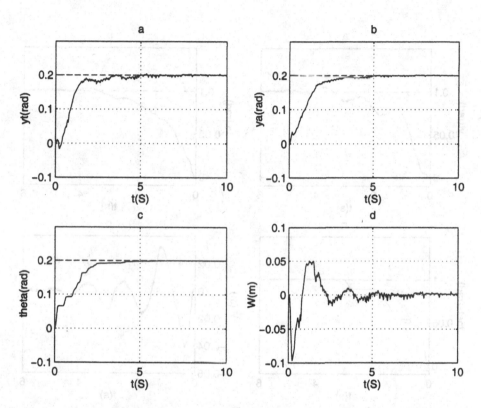

Figure 5.8. System responses to a 0.2 *rad* step input using the IDML neural network controller for $M_l = 850$ *g* ($\alpha = 0.48$ with modified learning rule): (a) actual tip position, (b) redefined output (c) hub position, and (d) tip deflection (dashed lines correspond to the reference trajectories).

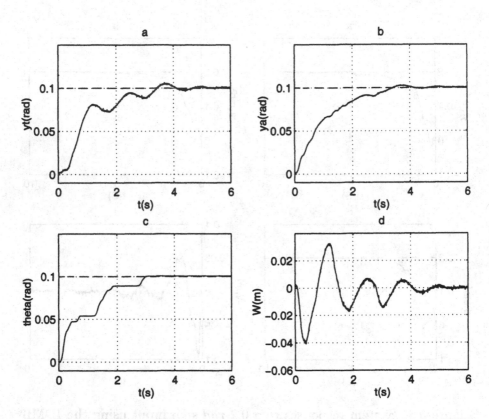

Figure 5.9. System responses to a 0.1 *rad* step input using the NRL neural network controller ($\alpha = 0.48$): (a) actual tip position, (b) redefined output (c) hub position, and (d) tip deflection (dashed lines correspond to the reference trajectories).

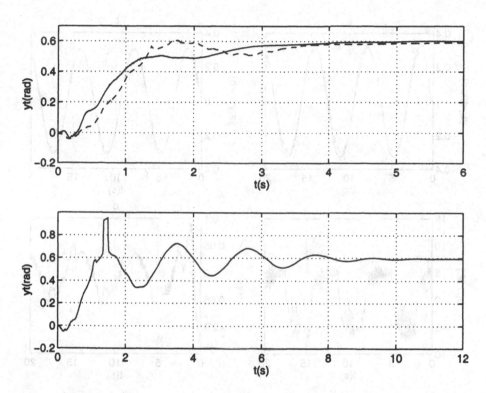

Figure 5.10. Actual tip responses to a 0.6 *rad* step input using the NRL neural network controller for different outputs: top- $\alpha = 0.48$ (solid line) and $\alpha = 0$ (dashed line) with modified learning rule; bottom- $\alpha = 0$, with no modifications.

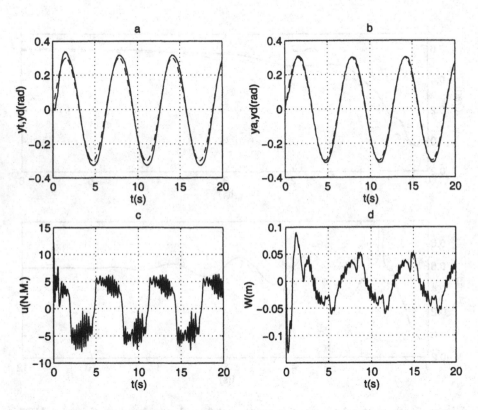

Figure 5.11. System responses to a 0.3 $sin(t)$ input using the NRL neural network controller ($\alpha = 0.48$): (a) actual tip position, (b) redefined output (c) control torque, and (d) tip deflection (dashed lines correspond to the desired trajectories).

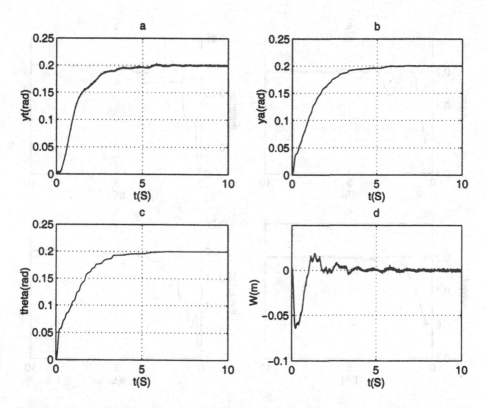

Figure 5.12. System responses to a 0.2 *rad* step input using the NRL neural network controller for $M_l = 850$ *g* ($\alpha = 0.48$ with modified learning rule): (a) actual tip position, (b) redefined output (c) hub position, and (d) tip deflection (dashed lines correspond to the reference trajectories).

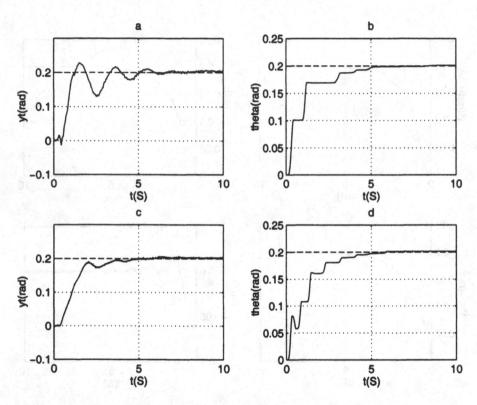

Figure 5.13. System responses to a 0.2 *rad* step input using the Joint-based neural network controller: (a) actual tip position (no modification), (b) hub position (no modification), (c)actual tip position (modified learning rule) and (d) hub position (modified learning rule) (dashed lines correspond to the reference trajectories).

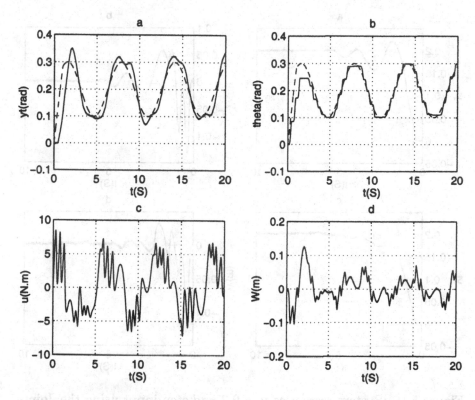

Figure 5.14. System responses to a $0.2 + 0.1$ $sin(t)$ input using the Joint–based neural network controller: (a) actual tip position, (b) Hub position (c) control torque, and (d) tip deflection (dashed lines correspond to the desired trajectories).

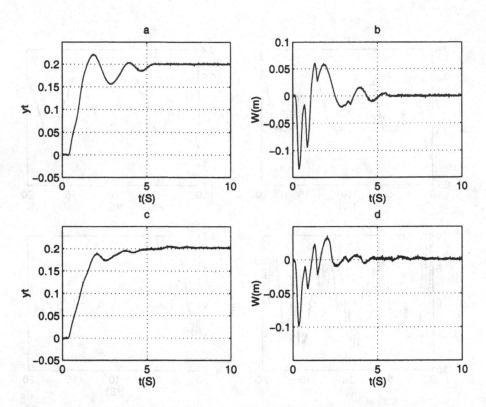

Figure 5.15. System responses to a 0.2 *rad* step input using the Joint-based neural network controller for $M_l = 850$ *g*: (a) actual tip position (No modifications), (b) total deflection (No modifications), (c) actual tip position (with deflection control) and (d) total deflection (with deflection control).

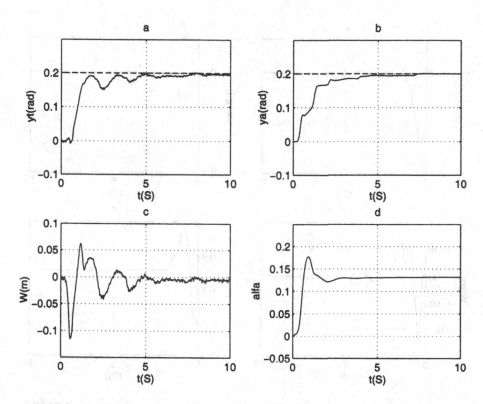

Figure 5.16. System responses to a 0.2 *rad* step input using the ORTOL neural network controller: (a) actual tip position, (b) redefined output (c) tip deflection, and (d) evolution of α. (dashed lines correspond to the reference trajectories).

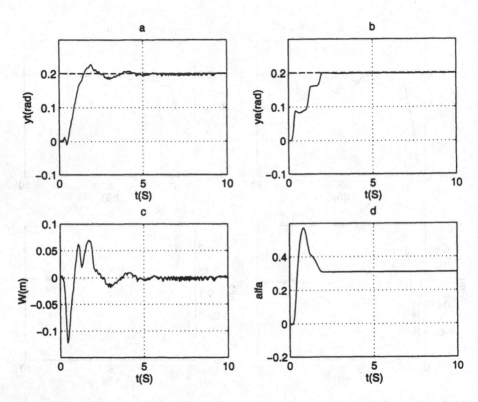

Figure 5.17. System responses to a 0.2 *rad* step input using the ORTOL neural network controller for $M_l = 850$ *g*: (a) actual tip position, (b) redefined output (c) tip deflection, and (d) evolution of α. (dashed lines correspond to the reference trajectories).

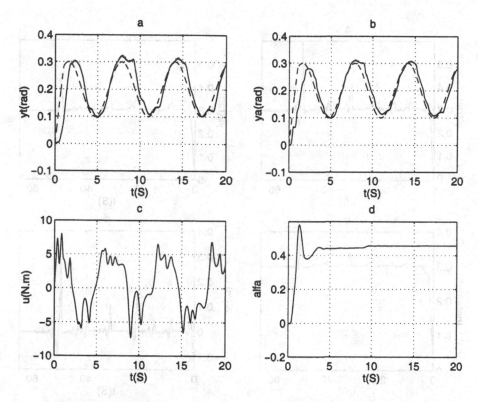

Figure 5.18. System responses to a $0.2 + 0.1 \, sin(t)$ input using the OR-TOL neural network controller: (a) actual tip position, (b) redefined output (c) control torque, and (d) tip deflection (dashed lines correspond to the desired trajectories).

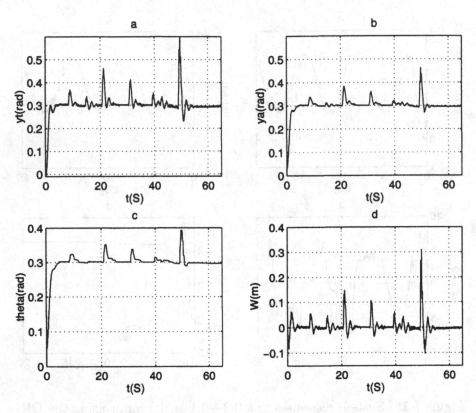

Figure 5.19. System responses to a 0.3 *rad* step input using the NRL neural network controller: a) actual tip position, (b) redefined output (c) hub position, and (d) tip deflection

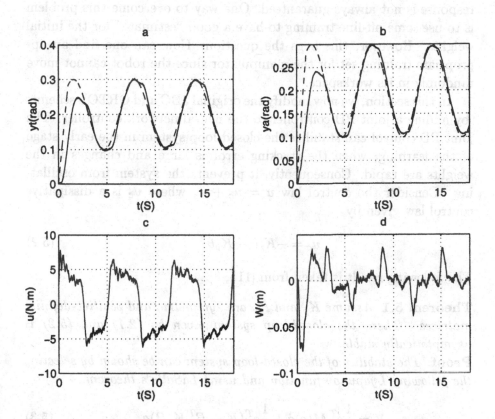

Figure 5.20. System responses to a $0.2 + 0.1\ sin(t)$ input using the NRL neural neural network controller for $M_l = 850\ g$ ($\alpha = 0.65$): (a) actual tip position, (b) redefined output (c) control torque, and (d) tip deflection (dashed lines correspond to the desired trajectories).

5.4.   Modified Schemes

As stated earlier, the JBC and ORTOL schemes assume no *a priori* knowledge about the system dynamics and no off-line training is performed. Also, experimental results show that when no *a priori* knowledge about the system dynamics is assumed, in general smooth transient response is not always guaranteed. One way to overcome this problem is to use some off-line training to have a good "estimate" for the initial weights. However, this begs the question: How can one find an appropriate training set for the manipulator since the robot cannot move randomly in its workspace.

In this section, we now modify the original JBC and ORTOL schemes by adding a joint PD controller to the neural network controller. The joint PD control can stabilize the closed-loop system in the early stage of the learning, when the tracking error is large and changes in the weights are rapid. Consequently, it prevents the system from oscillating. Consider the control law $u = u_c + u_n$ where u_c is a dissipative control law given by

$$u_c = -K_p\theta - K_v\dot{\theta} \qquad (5.2)$$

The following result is taken from [115, 8].

Theorem 5.1 *Assume K_p and K_v are symmetric and positive-definite matrices. Then, the closed-loop system given by (2.1) and (5.2) is asymptotically stable.*
Proof: *The stability of the closed-loop system can be shown by selecting the following Lyapunov function and using LaSalle's theorem.*

$$V = \frac{1}{2}\dot{q}^T M(q)\dot{q} + \frac{1}{2}q^T(K + B^T K_p B)q, \qquad (5.3)$$

where $B = [I_{n\times n} 0_{n\times m}]$. For details refer to [115, 8].

Since the result of this theorem is valid for any positive definite K_p and K_v, without using any *a priori* knowledge about the system dynamics, the control signal u_c defined by (5.2) can be added to the output of the neural network controller u_n to guarantee a stable closed-loop system in the initial stage of the learning period. This will also improve the transient response of the system. Experimental results show that significant improvement in the system response can be obtained by using this modification.

5.4.1. The Modified JBC Scheme

The modified structure is shown in Figure 5.21. In this scheme, a joint

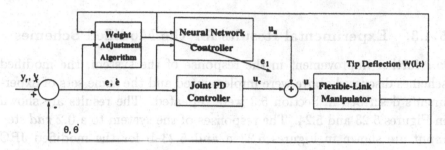

Figure 5.21. The modified Joint–based neural network controller.

PD controller and a neural network feedback controller are employed. The objective function for training the neural network is considered as $J = \frac{1}{2}(e^T K_1 e + \dot{e}^T K_2 \dot{e} + W(l,t)^T K_3 W(l,t))$, where $e := y_r - \theta$. The inputs to the network are e, $\dot{e}$, and $W(l,t)$, and the output of the network is u_n. Note that the structure of the neural network remains the same and the weights of the neural network can be adjusted in a similar manner to that described in Section 4.4.

5.4.2. The Modified ORTOL Scheme

The structure of the modified ORTOL scheme is shown in Figure 5.22. The joint PD control serves the same purpose as in the modified JBC

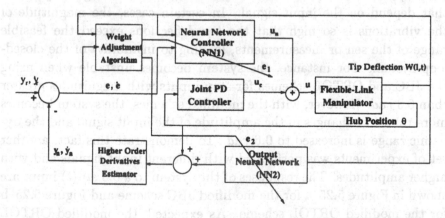

Figure 5.22. Structure of the modified neural network based controller using output re-definition.

case. Note that the structures of the neural networks remain the same and the weights of the neural network controller (NN1) can be adjusted in a similar manner to that described in Section 4.5.

5.4.3.   Experimental Results for the Modified Schemes

For further improvement in the response of the system, the modified schemes discussed above were implemented and the same sets of experiments described in Section 5.3 were repeated. The results are shown in Figures 5.23 and 5.24. The responses of the system to a 0.2 rad step input are shown in Figures 5.23–a and 5.23–b for the modified JBC scheme with $M_l = 30$ g, Figures 5.24–a and 5.24–b for the modified ORTOL scheme with $M_l = 30$ g, Figures 5.23–c and 5.23–d for the modified JBC scheme with $M_l = 850$ g, and Figures 5.24–c and 5.24–d for the modified ORTOL scheme with $M_l = 850$ g. The responses of the system to a $0.2 + 0.1 sin(t)$ input are shown in Figures 5.23–e and 5.23–f for the modified JBC scheme and Figures 5.24–e and 5.24–f for the modified ORTOL scheme. These responses show significant improvement when compared with Figures 5.14 and 5.18. More specifically, the transient and steady-state behavior of the system are improved and the vibration of the second flexible mode of the system (19 rad/s) which exists in Figures 5.14 and 5.18, is damped by using this modification.

One of the limitations of the JBC and ORTOL schemes is their robustness to changes in the magnitude of the desired trajectory. Since these two schemes assume no *a priori* knowledge about the system dynamics, the flexible modes of the system vibrate with high amplitudes that depend on the input signal. In certain cases, the magnitude of the vibrations is so high that the tip deflections exceed the feasible range of the sensor measurements, leading to instability of the closed-loop system. For instance, the system becomes unstable when using the JBC and ORTOL schemes for step inputs with magnitudes greater than 0.3 rad. However, with the modified schemes, the system becomes more robust to changes in the amplitude of the input signal and the dynamic range is increased to 0.6 rad. To demonstrate this fact, another set of experiments was performed with different input signals and with higher amplitudes. The responses of the system to a 0.5 $sin(t)$ input are shown in Figure 5.25–a for the modified JBC scheme and Figure 5.25–b for the modified ORTOL scheme. As expected, the modified ORTOL scheme exhibits less tracking error than the modified JBC scheme. The reason is that in the ORTOL scheme, the control is based on tracking a point closer to the tip, while in the JBC scheme the control is based on tracking the joint although the tip deflections are damped out. Figures 5.25–c and 5.25–d show the responses of the system to a 0.5 *rad* step input for the modified JBC and ORTOL schemes. As can be seen, the modified schemes are more robust to the input signal variations.

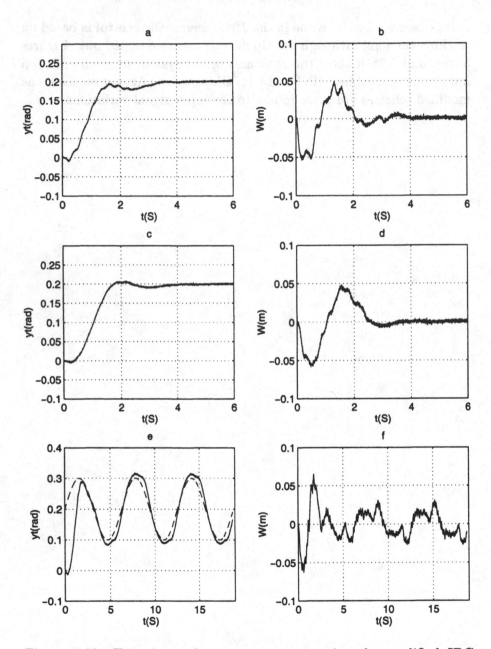

Figure 5.23. Experimental system responses using the modified JBC scheme for different payloads. For 0.2 *rad* step inputs, tip position: (a) $M_l = 30$ *g*, (c) $M_l = 850$ *g*; tip deflection: (b) $M_l = 30$ *g*, (d) $M_l = 850$ *g*. For a $0.2 + 0.1$ *sin(t)* input, tip position: (e), and tip deflection: (f) (dashed lines correspond to the reference trajectories).

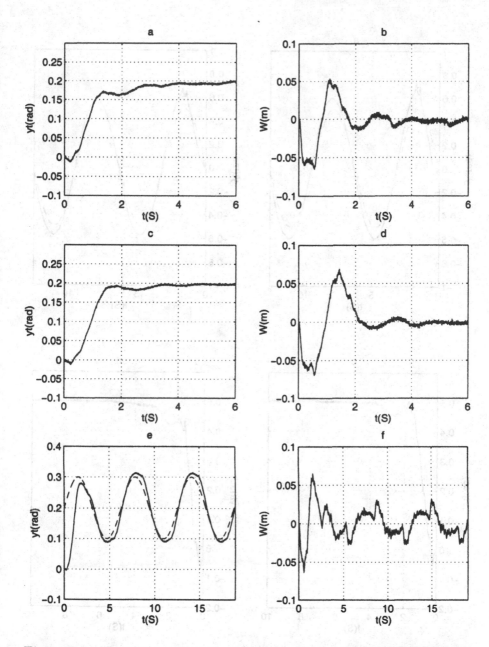

Figure 5.24. Experimental system responses using the modified ORTOL scheme for different payloads. For 0.2 *rad* step inputs, tip position: (a) $M_l = 30$ *g*, (c) $M_l = 850$ *g*; tip deflection: (b) $M_l = 30$ *g*, (d) $M_l = 850$ *g*. For a $0.2 + 0.1$ *sin(t)* input, tip position: (e), and tip deflection: (f) (dashed lines correspond to the reference trajectories).

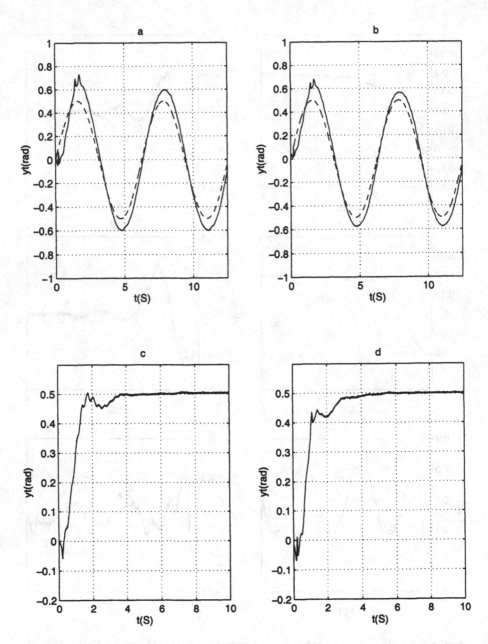

Figure 5.25. Experimental system responses to 0.5 $sin(t)$ and 0.5 rad inputs with (a,c) the modified JBC neural controller, (b,d) the modified ORTOL neural controller (dashed lines correspond to the reference trajectories).

Appendix A

The Viscous Damping Matrix

There has been some effort in the literature to model the viscous damping matrix F_2 introduced in (2.12). For instance, Moallem $et\ al.$ [52] used the $Rayleigh\ Damping$ method introduced in [88]. In [48], the authors used $f_i = a\sqrt{k_i}$, where f_i and k_i are the diagonal elements of the matrices F_2 and K respectively.

However, an exact model for F_2 is rarely known due to the uncertainty that is always present in the system. Therefore, some effort is required in selecting a value of α for redefinition of the output. By considering F_2 as an uncertain matrix, the variations of the roots of the characteristic polynomial of the matrix $A(\Upsilon)$ can be investigated due to changes in F_2. Several methods, mostly inspired by Kharitonov's result [116], have been developed in the literature for investigating the behavior of the roots of a polynomial with respect to parametric variations. One of the most relevant results deals with coefficients which can either be independently perturbed or in which perturbed parameters enter multilinearly [117, 118]. However, this result cannot be applied in a straightforward way since the coefficients of the characteristic polynomial of $A(\Upsilon)$ do not vary independently. In fact, $\alpha_i's$ appear nonlinearly in the polynomial coefficients. Nevertheless, use of both numerical and graphical techniques will be of great use.

Towards addressing the above problems, let us analyze the single flexible–link arm whose parameters are given in Table 2.1. Two flexible modes are considered for this study. First using the parameters given in [53] namely, $f_1 = 0.4$, $f_2 = 4$, the value of α^* is found to be 0.75 such that the matrix $A(\alpha)$ is Hurwitz. However, using numerical simulations we have found that this choice of α^* is not robust to variations in f_1 and f_2 (*i.e*, $f_1 = 0.4$ and $f_2 < 3.06$). Therefore, by considering f_1 and f_2 as unknown parameters, the characteristic polynomial of $A(\alpha)$ was computed using MAPLE [119]. The Routh–Hurwitz table was then constructed to determine the conditions under which the characteristic polynomial remains Hurwitz. We have found that a value of $\alpha^* = 0.6$ yields a matrix $A(\alpha)$ that is robust to variations of parameters f_1 and f_2 ranging from 1×10^{-8} to 10.

Using the α^* obtained for the zero payload mass ensures stability of the zero dynamics of the flexible manipulator when a nonzero payload mass is included in the model. Specifically, using the Routh–Hurwitz criterion, the condition under which the system is non–minimum phase was obtained using MAPLE [119]. It was found that the system is non–minimum phase only for a "negative" payload mass, M_l. Consequently, the value of α obtained for $M_l = 0$ ensures stability of the zero dynamics when one includes $M_l > 0$ in the system.

Appendix B

Numerical Models

B.1. Numerical Model for System I

The numerical model used in the simulations for System I is derived by using the assumed modes method with clamped–free shape functions. The link parameters are taken from [31] and are given in Table B.1.

The mass matrix M, the stiffness matrix K, and Coriolis and centrifugal terms obtained by using MAPLE [119] for two flexible modes are as follows (see Chapter 2 for the definition of the terms)

$$M(\delta) = \begin{bmatrix} m(\delta) & 0.3476 & -0.1880 \\ 0.3476 & 0.5648 & -0.4850 \\ -0.1880 & -0.4850 & 1.5332 \end{bmatrix},$$

$$K = \begin{bmatrix} 80.6190 & 0 \\ 0 & 3166 \end{bmatrix},$$

$$h_1(\dot{\theta}, \delta, \dot{\delta}) = 0.36\dot{\theta}[(\delta 1 - \delta 2)\dot{\delta}1 - (\delta 1 - \delta 2)\dot{\delta}2],$$

$$h_2(\dot{\theta}, \delta) = \begin{bmatrix} -0.18\dot{\theta}^2(\delta_1 - \delta_2) \\ -0.18\dot{\theta}^2(\delta_2 - \delta_1) \end{bmatrix}$$

where $m(\delta) = 1.5708 + 0.18(\delta_1^2 + \delta_2^2) - 0.36\delta_1\delta_2$.

B.2. Numerical Model for System II

The numerical model used in the simulations for System II (experimental manipulator) is derived by using the assumed modes method with clamped–free shape functions. The link parameters are given in Table 2.1.

Table B.1. Link parameters for System I

l	1.22 m
γ	0.24 kg/m
I_h	1.35 $kg.m^2$
EI	11.82 $N.m^2$
ω_1	16.5 rad/s
ω_2	103 rad/s
M_l	45 g

The mass matrix M, the stiffness matrix K, and Coriolis and centrifugal terms for two flexible modes are as follows

$$M(\delta) = \begin{bmatrix} m(\delta) & 1.0703 & -0.0282 \\ 1.0703 & 1.6235 & -0.4241 \\ -0.0282 & -0.4241 & 2.5920 \end{bmatrix},$$

$$K = \begin{bmatrix} 17.4561 & 0 \\ 0 & 685.5706 \end{bmatrix},$$

$$h_1(\dot{\theta}, \delta, \dot{\delta}) = 0.24\dot{\theta}[(\delta 1 - \delta 2)\dot{\delta} 1 - (\delta 1 - \delta 2)\dot{\delta} 2],$$

$$h_2(\dot{\theta}, \delta) = \begin{bmatrix} -0.12\dot{\theta}^2(\delta_1 - \delta_2) \\ -0.12\dot{\theta}^2(\delta_2 - \delta_1) \end{bmatrix}$$

where $m(\delta) = 0.9929 + 0.12(\delta_1^2 + \delta_2^2) - 0.24\delta_1\delta_2$. For Coulomb friction at the hub f_c, the hard nonlinearity $f_c = C_{coul}SGN(\dot{\theta})$ and its approximation $f_c = C_{coul}(\frac{2}{1+e^{-10\dot{\theta}}} - 1)$ were used and similar results were obtained.

B.3.   Two–link Manipulator

The dynamic model used in the simulations is derived based on the assumed modes method with clamped–mass shape functions given by

$$\phi_i(x) = cosh(\beta_i x/l_2) - cos(\beta_i x/l_2) - \gamma_i(sinh(\beta_i x/l_2) - sin(\beta_i x/l_2))$$

where l_2 is the length of the flexible link, x is the position variable along the flexible link, and the β_i's are obtained from

$$1 + cosh(\beta_i)cos(\beta_i) + \frac{M_l}{m}\beta_i(sinh(\beta_i)cos(\beta_i) - cosh(\beta_i)sin(\beta_i)) = 0$$

where $m = 0.210$ kg is the mass of the second link, and $M_l = 0.251$ kg is the payload mass. The first three β_i's are: 1.2030, 4.0159, and 7.1243.

The elements of the mass and stiffness matrices and Coriolis and centrifugal terms obtained by using MAPLE [119] for two flexible modes are as follows

$$
\begin{aligned}
M(1,1) &= m110 + co(m111 + m112\delta_1 + m113\delta_2) \\
&+ si(m114\delta_1 + m115\delta_2) + m116\delta_1^2 + m117\delta_2^2 \\
&+ m118\delta_1\delta_2 \\
M(1,2) &= m120 + co(m121 + m122\delta_1 + m123\delta_2) \\
&+ si(m124\delta_1 + m125\delta_2) + m126\delta_1^2 + m127\delta_2^2 \\
&+ m128\delta_1\delta_2 \\
M(1,3) &= m130 + m131co \\
M(1,4) &= m140 + m141co \\
M(2,2) &= m220 + m221\delta_1^2 + m222\delta_2^2 + m223\delta_1\delta_2 \\
M(2,3) &= m230 \quad M(2,4) = m240 \\
M(3,3) &= m330 \quad M(3,4) = m340 \\
M(4,4) &= m440
\end{aligned}
$$

$$
\begin{aligned}
f_1(1) + h_1(1) &= m113co\dot{\delta}_2\dot{\theta}_1 - m121si\dot{\theta}_2^{\,2} + m115si\dot{\delta}_2\dot{\theta}_1 \\
&+ m112co\dot{\delta}_1\dot{\theta}_1 + m114si\dot{\delta}_1\dot{\theta}_1 + 2m116\dot{\delta}_1\dot{\theta}_1\delta_1 \\
&+ m118\dot{\delta}_1\dot{\theta}_1\delta_2 + m122co\dot{\delta}_1\dot{\theta}_2 + m124si\dot{\delta}_1\dot{\theta}_2 \\
&+ 2m126\delta_1\dot{\delta}_1\dot{\theta}_2 + m128\delta_2\dot{\delta}_1\dot{\theta}_2 + 2m117\dot{\theta}_2\dot{\delta}_2\dot{\theta}_1 \\
&+ m118\dot{\delta}_1\dot{\delta}_2\dot{\theta}_1 + m123co\dot{\delta}_2\dot{\theta}_2 + m125si\dot{\delta}_2\dot{\theta}_2 \\
&+ 2m127\delta_2\dot{\delta}_2\dot{\theta}_2 + m128\delta_1\dot{\delta}_2\dot{\theta}_2 - m122si\dot{\theta}_2^{\,2}\delta_1 \\
&- m123\delta_2si\dot{\theta}_2^{\,2} + m124\delta_1co\dot{\theta}_2^{\,2} + m125\delta_2co\dot{\theta}_2^{\,2} \\
&- m111si\dot{\theta}_1\dot{\theta}_2\delta_1 - m113si\dot{\theta}_1\dot{\theta}_2\delta_2 + m114co\dot{\theta}_1\dot{\theta}_2\delta_1 \\
&+ m115co\dot{\theta}_1\dot{\theta}_2\delta_2 - m131\dot{\theta}_2si\dot{\delta}_1 - m141\dot{\theta}_2si\dot{\delta}_2
\end{aligned}
$$

$$
\begin{aligned}
f_1(2) + h_1(2) &= 0.5m111si\dot{\theta}_1^{\,2} + m131si\dot{\theta}_1\dot{\delta}_1 + m141si\dot{\theta}_1\dot{\delta}_2 \\
&+ 0.5m112si\dot{\theta}_1^{\,2}\delta_1 + 0.5m113si\dot{\theta}_1^{\,2}\delta_2 - 0.5m114co\dot{\theta}_1^{\,2}\delta_1 \\
&- 0.5m115co\dot{\theta}_1^{\,2}\delta_2 + m122co\dot{\delta}_1\dot{\theta}_1 + m124si\dot{\delta}_1\dot{\theta}_1 \\
&+ 2m126\delta_1\dot{\delta}_1\dot{\theta}_1 + m128\delta_2\dot{\delta}_1\dot{\theta}_1 + 2m221\delta_1\dot{\delta}_1\dot{\theta}_2 \\
&+ m223\delta_2\dot{\delta}_1\dot{\theta}_2 + m123co\dot{\delta}_2\dot{\theta}_1 + m125si\dot{\delta}_2\dot{\theta}_1 \\
&+ 2m127\delta_2\dot{\delta}_2\dot{\theta}_1 + m128\delta_1\dot{\delta}_2\dot{\theta}_1 + 2m222\delta_2\dot{\delta}_2\dot{\theta}_2
\end{aligned}
$$

$$+ \quad m223\dot{\delta}_2\dot{\theta}_2\delta_1$$

$$
\begin{aligned}
f_2(1) + h_2(1) \;=\; & -m131\dot{\theta}_2 si\dot{\theta}_1 - 0.5m112co\dot{\theta}_1{}^2 - 0.5m114si\dot{\theta}_1{}^2 \\
& - \; m116\dot{\theta}_1{}^2\delta_1 - 0.5m118\dot{\theta}_1{}^2\delta_2 - m122co\dot{\theta}_1\dot{\theta}_2 \\
& - \; m124si\dot{\theta}_1\dot{\theta}_2 - 2m126\delta_1\dot{\theta}_1\dot{\theta}_2 - m128\delta_2\dot{\theta}_1\dot{\theta}_2 \\
& - \; m221\dot{\theta}_2{}^2\delta_1 - 0.5m223\dot{\theta}_2{}^2\delta_2
\end{aligned}
$$

$$
\begin{aligned}
f_2(2) + h_2(2) \;=\; & -m141\dot{\theta}_2 si\dot{\theta}_1 - 0.5m113co\dot{\theta}_1{}^2 - 0.5m115si\dot{\theta}_1{}^2 \\
& - \; m117\dot{\theta}_1{}^2\delta_2 - 0.5m118\dot{\theta}_1{}^2\delta_1 - m123co\dot{\theta}_1\dot{\theta}_2 \\
& - \; m125si\dot{\theta}_1\dot{\theta}_2 - 2m127\delta_2\dot{\theta}_1\dot{\theta}_2 - m128\delta_1\dot{\theta}_1\dot{\theta}_2 \\
& - \; m222\dot{\theta}_2{}^2\delta_2 - 0.5m223\dot{\theta}_2{}^2\delta_1
\end{aligned}
$$

$$
K \;=\; \begin{bmatrix} 38.38 & 0 \\ 0 & 1506 \end{bmatrix}. \tag{B.1}
$$

where $M(i,j)$ represents the $(i,j)th$ element of the mass matrix and $f_i(j) + h_i(j)$ represents element j of the ith Coriolis and centrifugal terms.

The numerical values of the parameters in (B.1) are given below as

$$
\begin{aligned}
m110 \;&=\; 0.2176, \quad m111 = 0.1053, \quad m112 = 0, \quad m113 = 0, \\
m114 \;&=\; -0.3103, \quad m115 = 0.1936, \quad m116 = 1.2087, \\
m117 \;&=\; 1.2086, \quad m118 = -2.0079, \quad m120 = 0.1256, \\
m121 \;&=\; 0.0527, \quad m122 = 0, \quad m123 = 0, \quad m124 = -0.1552, \\
m125 \;&=\; 0.9679, \quad m126 = 1.2087, \quad m127 = 1.2086, \\
m128 \;&=\; -2.0079, \quad m130 = 0.3866, \quad m131 = 0.1552, \\
m140 \;&=\; -0.2969, \quad m141 = -0.0968, \quad m220 = 0.1256, \\
m221 \;&=\; 1.2087, \quad m222 = 1.2086, \quad m223 = -2.0079, \\
m230 \;&=\; 0.3866, \quad m240 = -0.2969, \quad m330 = 1.2090, \\
m340 \;&=\; -1.0048, \quad m440 = 1.2117
\end{aligned}
$$

Appendix C

Experimental Test–bed

Figure C.1 shows a schematic diagram of the test–bed and control system. The principal components of the system are as follows:

C.1. Spectrum TMS320C30 System Board

This board is employed to implement the control algorithms. It contains a TMS320C30 Digital Signal Processor (DSP) chip that contains integer and floating–point arithmetic units, 2048×32 bit words of on–chip RAM, 4096×32 bit words of on–chip ROM, control unit and parallel and serial interfaces. Operating from a 33.3 MHz clock, a performance of 16.7 million instructions per second is achieved.

The board occupies a single 16–bit slot within the PC host computer and is equipped with 64K words dual–port RAM which is intended for transfer of data between the PC and the DSP.

TMS320C30 system board also contains an analog I/O subsystem. There are two separate sets of analog-to-digital (A/D) converters, digital-to-analog (D/A) converters, and analog filters on input and output. The A/Ds are Burr-Brown PCM787 devices which offer 16–bit precision with up to 200 KHz sample rates. The sample/hold amplifiers SHC5320 and D/A converters PCM56P are also Burr Brown devices and are matched to the capabilities of the A/D. A ± 3 volt analog input range provides full scale operation of the A/D.

The TMS320C30 DSP board is equipped with the DSPLINK digital system expansion interface. This expansion bus is connected to a separate interface board that contains programmable timing circuitry and high current sources that control the pulsed current which drives the

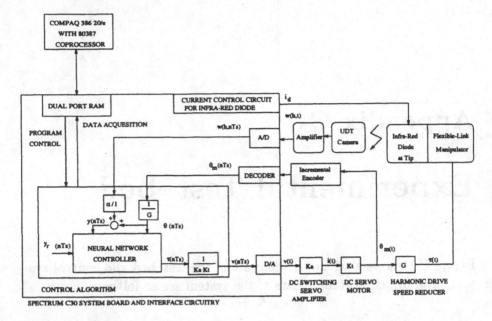

Figure C.1. Schematic of the experimental test–bed.

infrared diode mounted at the manipulator's tip. The interface board also includes decoder circuitry that decodes motor position information.

C.2.   The PC Host Computer

The host computer is a platform for the C30 system board. The user can control the program implemented on the C30 system board by sending different flags to the dual–port RAM that can be read by the DSP. Experimental data is transferred from the C30 system board to the dual–port RAM. The PC then reads the data and stores it for subsequent analysis.

C.3.   PWM Servo Amplifier

As Figure C.1 shows, a constant current $i(t)$ needs to be delivered to the motor armature. The Copley Control Corp. Model 215 is a transconductance pulse-width modulated (PWM) servo amplifier designed to drive the DC motors. It receives the D/A voltage $v(t)$ and delivers the output current $i(t) = K_a v(t)$. An internal control loop senses the output current and maintains K_a at the fixed value of 2.0. Consequently, the full scale D/A voltage range of ± 3 volts results in a maximum range of ± 6 amperes. The amplifier is switched with the rates of 22 to 26 KHz to eliminate audible noise. The 3–db bandwidth of the amplifier is 1 KHz.

C.4. DC Servo Motor

The EG&G Torque Systems Model MH3310–055G1 permanent magnet, brush type DC servo motor develops a constant torque $\tau(t) = K_t i(t)$, where the torque constant $k_t = 0.1175$ N.m/A. The armature current range of ± 6 amperes results in a maximum torque range of ± 0.705 N.m.

C.5. Harmonic Drive Speed Reducer

A harmonic drive is a high-ratio torque transmission device with almost zero backlash which is used in many electrically actuated robot manipulators, since DC motors are high–speed and relatively low–torque actuators. The HD systems Inc. Model RH20–CC harmonic drive transmits the torque generated by the DC motor to the manipulator's hub. A gear ratio of 50 : 1 is obtained. The speed reducer amplifies the motor torque by a factor of 50 and yields an output torque range of ± 35.25 N.m.

C.6. Incremental Encoder

Incremental encoders create a series of square waves. The number of square waves corresponds to the shaft revolutions. The Motion Control Devices Inc. Model M21 is a quadrature encoder that uses two output channels A & B in quadrature for position sensing. This allows the user to count the transitions and to view the state of the opposite channel during these transitions. Using this information, it can be determined if "A" leads "B" and thus the direction can be obtained. An incremental resolution of 500 cycles per revolutions of the motor shaft is obtained on each of two quadrature (A & B) signals.

C.7.   Quadrature Decoder

The Hewlett Packard HCTL–2020 consists of a $4X$ quadrature decoder, a binary up/down state counter, and an 8–bit bus interface. The quadrature decoder decodes the incoming signals from the encoder into count information. This circuitry multiplies the resolution of the input signals by a factor of four ($4X$ decoding). A 16–bit binary up/down position counter allows for software computation of absolute position.

A resolution of 2000 cycles per revolution of the motor shaft is obtained. This is equivalent to $50 \times 2000 = 100000$ cycles per revolution of the harmonic drive output shaft. However, the decoder resolution is 16 bits, corresponding to a maximum count of 65535 cycles. Consequently, the decoder can accommodate $\frac{65535}{100000} 2\pi = 1.31\pi$ radians of rotation before overflowing.

C.8.   Infrared Emitting Diode

Sensing the tip deflection needs a light source, a lens that focuses the image of the source onto a photodiode detector, a photodiode detector, and an amplifier to condition the detected signal. The positional resolution of the tip deflection is proportional to the signal to noise ratio (S/N) of the received signal at the output of the photodiode detector. The Opto Diode OD–50L Super High Power GaAlAs infrared emitting diode supplies up to 0.6 watts of peak optical power at a wavelength of 880 nm. An infrared light source, when used with a camera that incorporates a visible light blocking filter, reduces the interference from ambient light.

C.9.   UDT Camera

The United Detector Technology Model 274 camera consists of a wide angle lens and a SC–10D lateral–effect photodiode detector assembly. The 12.5 mm C-mount lens has a 55° field of view and includes a visible light blocking filter. The lens focuses the image of the infrared diode onto the photodiode detector. The diode appears as a light spot on the detector surface. The detector senses the centroid of the light spot and provides continuous analog output as the spot transverses the active area. An appropriate calibration procedure allows the user to calculate the absolute position of the infrared diode and hence the deflection of the tip.

C.10.   UDT Signal Conditioning Amplifier

The United Detector Technology Model 301DIV provides a transimpedance amplifier to condition the detector signals as well as the differential amplifiers necessary to generate a position related analog output. It interfaces the position–sensing photodiode detector to the A/D on the C30 system board. The amplifier is adjusted such that maximum tip deflections of ± 0.25 m correspond to output voltages of ± 3 volts. The 55° field of view of the lens should allow measurement of tip deflections in excess of ± 0.5 m. However, as the tip deflection increases, so does the magnitude of the slope of the arm evaluated at the tip, $\frac{\partial W(x,t)}{\partial x} \mid_{x=l}$. This slope causes a rotation of the diode when viewed from the camera's frame of reference [91]. This increasing slope, when coupled with the narrow bandwidth of the infrared diode results in a reduction of the optical power received at the lens. Beyond a deflection of ± 0.25 m, the diode cannot be accurately detected by the camera even though it is still within the lens's field of view.

Noise and nonlinearities within the photodiode detector result in a measurement error of ± 2.5 mm within a deflection range of ± 0.01 m. As the deflection increases to a maximum of ± 0.25 m, the error increases to ± 10 mm.

Bibliography

[1] A. Ficola, R. Marino, and S. Nicosia, "A singular perturbation approach to the control of elastic robots," in *Proceedings of the 21st Ann. Allerton Conf. Communication, Control, and Computing*, Univ. Illinois., 1983.

[2] L.M. Sweet and M.C. Good, "Re-definition of the robot motion control problem: Effects of plant dynamics, drive system constraints, and user requirements," in *Proceedings of the 23rd IEEE Conference on Decision and Control*, pp. 724–731, 1984.

[3] A.R. Fraser and R.W. Daniel, *Perturbation Techniques for Flexible Manipulators*. Kluwer Academic Publishers, 1991.

[4] D. Wang and M. Vidyasagar, "Feedback linearizability of multilink manipulators with one flexible link," in *Proceedings of the 28th IEEE Conference on Decision and Control*, pp. 2072–2077, 1989.

[5] D. Wang and M. Vidyasagar, "Control of a class of manipulators with a single link- part i: Feedback linearization," *ASME Journal of Dynamic Systems, Measurement, and Control*, pp. 655–661, December 1991.

[6] J.-J.E. Slotine and W. Li, *Applied Nonlinear Control*. Prentice-Hall, 1991.

[7] W.J. Book, O. Maizzo-Neto, and D.E. Whitney, "Feedback control of two beam, two joint systems with distributed flexibility," *ASME Journal of Dynamic Systems, Measurement and Control*, pp. 424–431, 1975.

[8] A. De Luca and B. Siciliano, "An asymptotically stable joint PD controller for robot arms with flexible links under gravity," in

Proceedings of the 31st IEEE Conference on Decision and Control, pp. 325–326, 1992.

[9] R.N. Banavar and P. Dominic, "An LQG/H_∞ controller for a flexible manipulator," *IEEE Trans. on Control Systems Technology*, Vol. 3, pp. 409–416, December 1995.

[10] P.T. Kotnik, S. Yurkovich, and U. Ozguner, "Acceleration feedback for control of a flexible manipulator arm," *J. of Robotic Systems, Vol. 5, No. 3*, pp. 181–196, 1988.

[11] S.S. Ge, T.H. Lee, and G. Zhu, "Improving joint PD control of single-link flexible robots by strain/tip feedback," in *Proceedings of the IEEE Int. Conf. on Control Applications*, pp. 965–969, 1996.

[12] S.S. Ge, T.H. Lee, and G. Zhu, "Tip tracking control of a flexible manipulator using PD type controller," in *Proceedings of the IEEE Int. Conf. on Control Applications*, pp. 309–313, 1996.

[13] R.H. Canon and J.E. Schmitz, "Initial experiments on the endpoint control of a flexible one-link robot," *International Journal of Robotics Research*, Vol. 3, No. 3, pp. 62–75, Fall 1984.

[14] Y. Sakawa, F. Matsumo, and S. Fukushima, "Modeling and feedback control of a flexible arm," *Journal of Robotic Systems*, Vol. 2, No. 4, pp. 453–472, 1985.

[15] G.G. Hastings and W.J. Book, ""Experiments in Optimal Control of a Flexible Arm"," in *Proceedings of the American Control Conference*, pp. 728–729, June 1985.

[16] P. Karkkainen, "Compensation of manipulator flexibility effects by modal space techniques," in *Proceedings of the IEEE Int. Conf. on Robotics and Automation*, 1985.

[17] E. Schmitz, *Experiments on the End-point Control of a very Flexible one-Link Manipulator*. PhD thesis, Stanford University, 1985.

[18] P.B. Usoro, R. Nadira, and S.S. Mahil, "Control of lightweight flexible manipulator arms: A feasibility study," tech. rep., N.S.F., 1986.

[19] E. Bayo, "A finite-element approach to control the end-point motion of a single-link flexible robot," *Journal of Robotic Systems*, pp. 63–75, 1987.

[20] O.S. Kwon and W.J. Book, "An inverse dynamic method yielding flexible manipulator state trajectories," in *Proceedings of the American Control Conference*, pp. 186–193, 1990.

[21] D.C. Nemir, A.J. Koivo and R.L. Kashyap, "Pseudolinks and the self-tuning control of a nonrigid link mechanism," *IEEE Trans. on Systems, Man, and Cybernetics*, pp. 40–48, January 1988.

[22] N.C. Singer and W.P. Seering, "Design and comparison of command shaping methods for controlling residual vibration," in *Proceedings of the IEEE Int. Conf. on Robotics and Automation*, pp. 888–893, 1989.

[23] N.C. Singer and W.P. Seering, "Experimental verification of command shaping methods for controlling residual vibration in flexible robots," in *Proceedings of the American Control Conference*, pp. 1738–1744, 1990.

[24] K.L. Hillsley and S. Yurkovich, "Vibration control of a two-link flexible robot arm," in *Proceedings of the IEEE Int. Conf. on Robotics and Automation*, pp. 2121–2126, 1991.

[25] D.P. Magee and W.J. Book, "Eliminating multiple modes of vibration in a flexible manipulator," in *Proceedings of the IEEE Int. Conf. on Robotics and Automation*, pp. 474–479, 1993.

[26] D.P. Magee and W.J. Book, "Implementing modified command filtering to eliminate multiple modes of vibration," in *Proceedings of the American Control Conference*, pp. 2700–2704, 1993.

[27] A. Tzes and S. Yurkovich, "An adaptive input shaping control scheme for vibration suppression in slewing flexible structures," *IEEE Trans. on Control System Technology*, pp. 114–121, June 1993.

[28] F. Khorrami, S. Jain, and A. Tzes, "Experiments on rigid body based controllers with input preshaping for a two-link flexible manipulator," in *Proceedings of the American Control Conference*, pp. 2957–2961, 1992.

[29] J.H. Chow and P.V. Kokotovic, "Two-time-scale feedback design of a class of nonlinear systems," *IEEE Trans. on Automatic Control*, pp. 438–443, June 1978.

[30] P.V. Kokotovic, H.K. Khalil, and J. O'Reilly, *Singular Perturbation Methods in Control: Analysis and Design*. Academic Press, 1986.

[31] B. Siciliano and W.J. Book, "A singular perturbation approach to control of lightweight flexible manipulators," *International Journal of Robotics Research*, pp. 79–90, August 1988.

[32] K. Khorasani and M.W. Spong, "Invariant manifolds and their application to robot manipulators with flexible joints," in *Proceedings of the IEEE Int. Conf. on Robotics and Automation*, pp. 978–983, 1985.

[33] Y. Aoustin and C. Chevallereau, "The singular perturbation control of a two-flexible-link robot," in *Proceedings of the IEEE Int. Conf. on Robotics and Automation*, pp. 737–742, 1993.

[34] F.L. Lewis and M. Vandergift, "Flexible robot arm control by a feedback linearization/ singular perturbation approach," in *Proceedings of the IEEE Int. Conf. on Robotics and Automation, Atlanta, Georgia*, pp. 729–736, 1993.

[35] D.A. Schoenwald and U. Ozguner, "On combining slewing and vibration control in flexible manipulator via singular perturbations," in *Proceedings of the 29th IEEE Conference on Decision and Control*, pp. 533–538, 1990.

[36] B. Siciliano, J.V.R. Prasad, and A.J. Calise, "Design of a composite controller for a two-link flexible manipulator," in *Proccedings of the Int. Symp. on Intelligent Robotics, Bangalore, India*, pp. 126–137, 1991.

[37] B. Siciliano, J.V.R. Prasad, and A.J. Calise, "Output feedback two–time scale control of multi–link flexible arms," *ASME Journal of Dynamic Systems, Measurement, and Control*, Vol. 114, pp. 70–77, 1992.

[38] Y. Aoustin, C. Chevallereau, A. Glumineau, and C.H. Moog, "Experimental results for the end–effector control of a single flexible

robotic arm," *IEEE Trans. on Control System Technology*, Vol. 2, pp. 371–381, 1994.

[39] M.W. Spong, K. Khorasani, and P.V. Kokotovic, "An integral manifold approach to the feedback control of flexible joint robots," *IEEE Journal of Robotics and Automation*, pp. 291–300, August 1987.

[40] K. Hashtrudi Zaad and K. Khorasani, "Control of nonminimum phase singularly perturbed systems with applications to flexible-link manipulators," in *Proceedings of the Workshop on Advances in Control and its Applications*, pp. 234–265, Lecture Notes in Control and Information Sciences 208, Springer-Verlag, 1995.

[41] M. Moallem, K. Khorasani, and R.V. Patel, "An integral manifold approach for tip position tracking of flexible multi-link manipulators," *IEEE Trans. on Robotics and Automation*, Vol. 13, No. 6, pp. 823–837, December 1997.

[42] B. Siciliano, W.J. Book, and G. De Maria, "An integral manifold approach to control of a one-link flexible arm," in *Proceedings of the 25th IEEE Conference on Decision and Control*, 1986.

[43] M. Moallem, R.V. Patel, and K. Khorasani, "Flexible–link robot manipulators: Control techniques and structural design." To be published in Lecture Notes in Control and Information Sciences, Springer-Verlag, Heidelberg, 2000.

[44] D. Wang and M. Vidyasagar, "Transfer functions for a single flexible link," in *Proceedings of the IEEE Int. Conf. on Robotics and Automation*, pp. 1042–1047, 1989.

[45] H.R. Pota and M. Vidyasagar, "Passivity of flexible beam transfer functions with modified outputs," in *Proceedings of the IEEE Int. Conf. on Robotics and Automation*, pp. 2826–2831, 1991.

[46] D. Wang and M. Vidyasagar, "Passive control of a stiff flexible link," *International Journal of Robotics Research*, pp. 572–578, December 1992.

[47] A. De Luca and L. Lanari, "Achiving minimum phase behavior in a one-link flexible arm," in *Proccedings of the Int. Symp. on Intelligent Robotics ,Bangalore, India*, pp. 224–235, 1991.

[48] A. De Luca and B. Siciliano, "Trajectory control of a nonlinear one-link flexible arm," *International Journal of Control,* Vol. 50, No. 5, pp. 1699–1715, 1989.

[49] S.K. Madhavan and S.N. Singh, "Inverse trajectory control and zero dynamic sensitivity of an elastic manipulator," *International Journal of Robotics and Automation,* pp. 179–191, 1991.

[50] C.I. Byrnes and A. Isidori, "Global feedback stabilization of nonlinear systems," in *Proceedings of the 24th IEEE Conference on Decision and Control,* pp. 1031–1037, 1985.

[51] R.M. Hirschorn, "Invertibility of nonlinear control systems," *SIAM Journal of Control and Optimization,* Vol. 17, No. 2, pp. 289–297, March 1979.

[52] M. Moallem, R.V. Patel, and K. Khorasani, "An Inverse Dynamics Control Strategy for Tip Position Tracking of Flexible Multi–Link Manipulators," *Journal of Robotic Systems,* Vol. 14, No. 9, pp. 649–658, 1997.

[53] H. Geniele, R.V. Patel, and K. Khorasani, "Control of a flexible-link manipulator," in *Proceedings of the IEEE Int. Conf. on Robotics and Automation,* pp. 1217–1222, 1995.

[54] H. Geniele, R.V. Patel, and K. Khorasani, "Control of a flexible-link manipulator," in *Proceedings of the Fourth Int. Symp. on Robotics and Manufacturing,* pp. 567–572, November 1992.

[55] H. Geniele, R.V. Patel, and K. Khorasani, "End-Point Control of a Flexible-Link Manipulator: Theory and Experiments," *IEEE Trans. on Control System Technology,* Vol. 5, No. 6, pp. 556–570, November 1997.

[56] P. Misra and R.V. Patel, "Transmission Zero Assignment in Linear Multivariable Systems Part I: Square Systems," in *Proceedings of the 27th IEEE Conference on Decision and Control,* pp. 1310–1311, 1988.

[57] R.V. Patel and P. Misra, "Transmission Zero Assignment in Linear Multivariable Systems Part II: The General Case," in *Proceedings of the American Control Conference,* pp. 644–648, 1992.

[58] S. Gopalswamy and J.K. Hedrick, "Tracking nonlinear nonminimum phase systems using sliding control," *International Journal of Control*, pp. 1141–1158, 1993.

[59] K.Y. Lian, L.C. Fu, and T.L. Liao, "Output tracking control of nonlinear systems with weakly nonminimum phase," in *Proceedings of the American Control Conference*, pp. 3081–3085, 1992.

[60] L. Benvenuti, M.D. Di Benedetto, and J.W. Grizzle, "Approximate output tracking for nonlinear non-minimum phase systems with an application to flight control," *Int. J. of Robust and Nonlinear Control*, Vol. 4, pp. 397–414, 1994.

[61] J. Hauser, S. Sastry, and G. Meyer, "Nonlinear control design for slightly nonminimum phase systems: Application to V/STOL aircraft," *Automatica*, Vol. 28, No. 4, pp. 665–679, July 1992.

[62] C.W. De Silva, "An analytical framework for knowledge–based tuning of servo controller," *Eng. App. Art. Intell., Vol. 4, No. 3*, pp. 177–189, 1991.

[63] A. Ollero and A. Garcia–Gerezo, "Direct digital control auto–tuning and supervision using fuzzy logic," *Fuzzy Sets & Systems*, pp. 135–153, 1989.

[64] S. Tzafestas and N. Papanikolopoulos, "Incremental fuzzy expert PID control," *IEEE Trans. on Ind. Electron.*, Vol. 37, pp. 365–371, October 1990.

[65] H. Van Nauta Lemke and W. De-zhao, "Fuzzy PID supervisor," in *Proceedings of the 24th IEEE Conference on Decision and Control*, pp. 602–608, 1985.

[66] E. Kubica and D. Wang, "A fuzzy control strategy for a flexible link robot," in *Proceedings of the IEEE Int. Conf. on Robotics and Automation*, pp. 236–241, 1993.

[67] V.G. Moudgal, W.A. Kwong, K.M. Passino, and S. Yurkovich, "Fuzzy learning control for a flexible–link robot," in *Proceedings of the American Control Conference*, pp. 563–567, June 1994.

[68] E. Garcia-Benitez, S. Yurkovich, and K.M. Passino, "A fuzzy supervisor for flexible manipulator control," in *IEEE Int. Symposium on Intelligent Control*, pp. 37–42, August 1991.

[69] E. Garcia-Benitez, S. Yurkovich, and K.M. Passino, "Rule-based supervisory control of a two-link flexible manipulator," *J. Intell. Robotic Syst.*, Vol. 7, pp. 195–213, 1993.

[70] V.G. Moudgal, K.M. Passino, and S. Yurkovich, "Rule–based control for a flexible–link robot," *IEEE Trans. on Control System Technology*, Vol. 2, No. 4, pp. 392–405, December 1994.

[71] C.L. McCullough, "Neural network control vs. anticipatory fuzzy control for a flexible beam: A comparison," in *Proceeedings of IEEE Int. Conf. on Neural Networks*, pp. 2350–2354, 1994.

[72] W. Cheng and J.T. Wen, "A neural controller for the tracking control of flexible arms," in *Proceedings of the IEEE Int. Conf. on Neural Networks*, pp. 749–754, 1993.

[73] N. Mahmood and B.L. Walcott, "Neural network based adaptive control of a flexible link manipulator," in *Proceedings of the IEEE National Aerospace and Electronics Conference*, pp. 851–857, 1993.

[74] J.D. Donne and U. Ozguner, "Neural control of a flexible-link manipulator," in *Proceedings of the IEEE Int. Conf. on Neural Networks*, pp. 2327–2332, 1994.

[75] Y. Iiguni, H. Sakai, and H. Tokumaru, "A nonlinear regulator design in the presence of system uncertainties using multilayered neural networks," *IEEE Trans. on Neural Networks, Vol. 2, No. 4*, pp. 410–417, July 1991.

[76] M.K. Sundareshan and C. Askew, "Neural network-based payload adaptive variable structure control of a flexible manipulator arm," in *Proceeedings of IEEE Int. Conf. on Neural Networks*, pp. 2616–2621, 1994.

[77] A. Register, W.J. Book, and C.O. Alford, "Artificial neural network control of a nonminimum phase, single-flexible-link," in *Proceedings of IEEE Int. Conf. on Robotics and Automation*, pp. 1935–1940, 1996, Vol. 2.

[78] K. Takahashi and I. Yamada, "Neural–network–based learning control of flexible mechanism with application to a single–link flexible arm," in *Proceedings of the ASME WAM Intelligent Control Systems*, vol. 48, pp. 95–104, 1993.

[79] R.T. Newton and Y. Xu, "Neural network control of a space manipulator," *IEEE Control Systems Magazine*, pp. 14–22, December 1993.

[80] A. Yesildirek, M.W. Vandergift, and F.L. Lewis, "A neural network controller for flexible-link robots," in *Proceedings of the IEEE Int. Symposium on Intelligent Control*, pp. 63–68, 1994.

[81] S. Megahed and M. Renaud, "Minimization of the computation time necessary for the dynamic control of robot manipulator," in *Proceedings of the 12th International Symposium on Industrial Robots*, pp. 469–478, 1982.

[82] M. Spong and M. Vidyasagar, *Robot Dynamics and Control*. McGraw Hill, 1989.

[83] L. Meirovitch, *Elements of Vibration Analysis*. New York: McGraw Hill, 1975.

[84] F. Bellezza, G. Lanari, and G. Ulivi, "Exact modeling of the flexible slewing link," in *Proceedings of the IEEE Int. Conf. on Robotics and Automation*, pp. 734–739, 1990.

[85] S. Cetinkunt and W.-L. Yu, "Closed-loop behavior of a feedback-controlled flexible arm : A comparative study," *International Journal of Robotics Research*, Vol. 10, No. 3, pp. 263–275, June 1991.

[86] G.G. Hastings and W.J. Book, "A linear dynamic model for flexible robotic manipulators," *IEEE Control Systems Magazine*, pp. 61–64, 1987.

[87] L. Meirovitch, *Anlytical Methods in Vibrations*. New York: Macmillan, 1967.

[88] W.T. Thomson, *Theory of Vibrations with Applications*. Englewood Cliffs, N.J.: Prentice-Hall, 3 ed., 1988.

[89] W.J. Book, "Recursive Lagrangian Dynamics of Flexible Manipulator Arms," *International Journal of Robotics Research*, Vol. 3, No. 3, pp. 87–100, Fall 1984.

[90] J. Hauser and R.M. Murray, "Nonlinear controllers for non-integrable systems: The acrobot example," in *Proceedings of the American Control Conference*, pp. 669–671, 1990.

[91] H. Geniele, "Control of a flexible-link manipulator," Master's thesis, Concordia University, Montreal, Canada, 1994.

[92] A. Tornambe, "Output feedback stabilization of a class of non-minimum phase nonlinear systems," *Systems and Control Letters*, Vol. 19, pp. 193–204, 1992.

[93] J.K. Hedrick and S. Gopalswamy, "Nonlinear flight control design via sliding methods," *J. Guidance*, Vol. 13, No. 5, pp. 850–858, September/October 1990.

[94] J.-H. Park and H. Asada, "Design and control of minimum-phase flexible arms with torque transmission mechanisms," in *Proceedings of the IEEE Int. Conf. on Robotics and Automation*, pp. 1790–1795, 1990.

[95] A. Isidori and C. Moog, "On the nonlinear equivalent of the notion of transmission zeros," in *Modeling and Adaptive Control*, Springer Verlag, 1987.

[96] G. Cybenko, "Approximation by superposition of sigmoidal functions," *Mathematics of Control, Signals and Systems,*, pp. 303–314, October 1989.

[97] K.I. Funahashi, "On the approximation of realization of continuous mappings by neural networks," *Neural Networks*, Vol.2, pp. 183–192, 1989.

[98] R. Hecht-Nielsen, "Theory of the backpropagation neural network," in *Proceedings of the Int. Joint Conf. on Neural Networks*, pp. 593–605, 1989.

[99] M. Stinchombe and H. White, "Approximating and learning unknown mapppings using multilayered feedforward networks with bounded weights," in *Proceedings of the Int. Joint Conf. on Neural Networks*, pp. 7–16, 1990.

[100] A.N. Kolmogorov, "On the representation of continuous functions of many variables by superposition of continuous functions of one

variable and addition," *Dokl. Akad. Nauk USSR, 114*, pp. 953–956, 1957.

[101] K.S. Narendra and S. Mukhopadhyay, "Adaptive control of non-linear multivariable systems using neural networks," *Neural Networks, Vol. 7, No. 5*, pp. 737–752, 1994.

[102] K.S. Narendra and S. Mukhopadhyay, "Neural networks in control systems," in *Proceedings of the 31st IEEE Conference on Decision and Control*, pp. 1–6, 1992.

[103] K.S. Narendra and K. Parthasarathy, "Identification and control of dynamical systems using neural networks," *IEEE Trans. on Neural Networks*, Vol. 1, No. 1, pp. 4–27, March 1990.

[104] G. Lightbody, Q.H. Wu, and G.W. Irwin, "Control application for feedforward networks," in *Neural Networks for Control* (Miller, T.W. *et al.*, ed.), pp. 51–71, MIT Press, Cambridge, MA., 1990.

[105] D. Psaltis, A. Sideris, and A.A. Yamamura, "A multilayered neural network controller," *IEEE Control Systems Magazine*, pp. 17–21, April 1988.

[106] M. Saerens and A. Soquet, "A neural controlle," in *Proceedings of the 1st IEE Conference on Neural Networks*, pp. 211–215, 1987.

[107] D.H. Nguyen and Widrow, B., "Neural networks for self-learning control systems," *IEEE Control Systems Magazine*, pp. 18–23, April 1990.

[108] M.I. Jordan and R.A. Jacobs, "Learning to control an unstable system with forward modeling," in *Advances in Neural Information Processing Systems*, vol. 2, pp. 324–331, Morgan Kaufman, San Mateo, CA., 1990.

[109] H. Miyamoto, M. Kawato, T. Setoyama, and R. Suzuki, "Feedback-error-learning neural network for trajectory control of a robotic manipulator," *Neural Networks, Vol. 1*, pp. 251–265, 1988.

[110] H. Gomi and M. Kawato, "Neural network control for a closed-loop system using feedback-error-learning," *Neural Networks*, Vol. 6, pp. 933–946, 1993.

[111] P.D. Wasserman, *Neural Computing: Theory and practice*. Van Nostrand Reinhold, New York, 1989.

[112] J.M. Zurada, *Introduction to Artificial Neural Systems*. West Publishing Company, 1992.

[113] H. Demuth and M. Beale, *Neural Network Toolbox User's Guide*. The MATH WORKS Inc., 1994.

[114] D.H. Nguyen and B. Widrow, "Improving the learning speed of 2-layer neural networks by choosing initial values of the adaptive weights," in *Proceedings of the Int. Joint Conf. on Neural Networks*, pp. 21–26, 1990.

[115] A.G. Kelkar, S.M. Joshi, and T.E. Alberts, "Globally stabilizing controllers for flexible multibody systems," in *Proceedings of the 31st IEEE Conference on Decision and Control*, pp. 2856–2859, 1992.

[116] V.L. Kharitonov, "Asymptotic stability of an equilibrium position of a family of systems of linear differential equations," *Differential'nye Uraveniya*, Vol. 14, No. 11, pp. 1483–1485, 1978.

[117] B.R. Barmish, "A generalization of Kharitonov's four-polynomial concept for robust stability problems with linearly dependent coefficient perturbations," *IEEE Transactions on Automatic Control*, Vol. 34, No. 2, pp. 157–164, 1990.

[118] A.C. Bartlett, C.V. Hollot, and H. Lin, "Root locations of an entire polytope of polynomials: It suffices to check the edges," *Math. Control, Signals, Systems*, Vol. 1, pp. 61–71, 1987.

[119] D. Redfern, *The MAPLE Handbook*. Springer Verlag, New York, 1993.

Index

Lecture Notes in Control and Information Sciences

Edited by M. Thoma and M. Morari

1997–2000 Published Titles: